# Dynamic Programming and Bayesian Inference, Concepts and Applications

# Dynamic Programming and Bayesian Inference, Concepts and Applications

*Editor*

Brygida Cullen

**Dynamic Programming and Bayesian Inference, Concepts and Applications**
Edited by **Brygida Cullen**

ISBN: 978-1-68117-200-2
Library of Congress Control Number: 2016934747

© 2017 by
SCITUS Academics LLC,
www.scitusacademics.com
Box No. 4766, 616 Corporate Way,
Suite 2, Valley Cottage,
NY 10989

This book contains information obtained from highly regarded resources. Copyright for individual articles remains with the authors as indicated. All chapters are distributed under the terms of the Creative Commons Attribution License, which permits unrestricted use, distribution, and reproduction in any medium, provided the original author and source are credited.

**Notice**

Reasonable efforts have been made to publish reliable data and views articulated in the chapters are those of the individual contributors, and not necessarily those of the editors or publishers. Editors or publishers are not responsible for the accuracy of the information in the published chapters or consequences of their use. The publisher believes no responsibility for any damage or grievance to the persons or property arising out of the use of any materials, instructions, methods or thoughts in the book. The editors and the publisher have attempted to trace the copyright holders of all material reproduced in this publication and apologize to copyright holders if permission has not been obtained. If any copyright holder has not been acknowledged, please write to us so we may rectify.

# PREFACE

A dynamic programming (DP) is an algorithmic technique which is usually based on a recurrent formula and one (or some) starting states. A sub-solution of the problem is constructed from previously found ones. Dynamic programming solutions have a polynomial complexity which assures a much faster running time than other techniques like backtracking, brute-force etc. Dynamic programming is both a mathematical optimization method and a computer programming method. In both contexts it refers to simplifying a complicated problem by breaking it down into simpler sub-problems in a recursive manner. While some decision problems cannot be taken apart this way, decisions that span several points in time do often break apart recursively. Bayesian inference is a method of statistical inference in which Bayes' theorem is used to update the probability for a hypothesis as more evidence or information becomes available. Dynamic programming algorithms are applied for optimization. A dynamic programming algorithm will inspect the previously solved sub-problems and will combine their solutions to give the best solution for the given problem. The alternatives are many, such as using a greedy algorithm, which picks the locally optimal choice at each branch in the road. The locally optimal choice may be a poor choice for the overall solution. While a greedy algorithm does not guarantee an optimal solution, it is often faster to calculate. Fortunately, some greedy algorithms are proven to lead to the optimal solution. Dynamic programming and Bayesian inference have been both intensively and extensively advanced in the course of recent years. As a consequence of these developments, interest in dynamic programming and Bayesian inference and their applications has greatly increased at all mathematical levels. This book, Dynamic programming and Bayesian inference, Concepts and Applications, is intended to provide some applications of Bayesian optimization and dynamic programming. This book presents a wide-ranging and demanding dealing of dynamic programming.

# CONTENTS

Chapter 1     Using Dynamic Programming Based on Bayesian Inference in Selection Problems ................................. 1

Chapter 2     Intelligent Condition Diagnosis Method Based on Adaptive Statistic Test Filter and Diagnostic Bayesian Network .................................................... 29

Chapter 3     Bayesian Markov Regime-Switching Models for Cointegration ..................................................... 63

Chapter 4     Classification of Web Services Using Bayesian Network ................................................................ 77

Chapter 5     Lane-Level Road Information Mining from Vehicle GPS Trajectories Based on Naïve Bayesian Classification .................... 93

Chapter 6     Semi-Supervised Bayesian Classification of Materials with Impact-Echo Signals ................... 129

Chapter 7     Optimal Allocation of Radio Resource in Cellular LTE Downlink Based on Truncated Dynamic Programming under Uncertainty ............ 165

Chapter 8     Variational Bayesian mixed-effects inference for classification studies ......................................... 191

Chapter 9  Structural learning of Bayesian networks
          by bacterial foraging optimization............................ 243

Index ................................................................................................ 295

# CHAPTER 1

## USING DYNAMIC PROGRAMMING BASED ON BAYESIAN INFERENCE IN SELECTION PROBLEMS

Mohammad Saber Fallah Nezhad[1]

[1] Associate Professor of Industrial Engineering, Yazd University, Iran

## 1. INTRODUCTION

An important subject in mathematical science that causes new improvements in data analysis is sequential analysis. In this type of analysis, the number of required observations is not fixed in advance, but is a variable and depends upon the values of the gathered observation. In sequential analysis, at any stage of data gathering process, to determine the number of required observations at the next stage, we analyze the data at hand and with respect to the obtained results, we determine how many more observations are necessary. In this way, the process of data gathering is cheaper and the information is used more effectively. In other words, the data gathering process in sequential analysis, in contrast to frequency analysis, is on-line. This idea caused some researches to conduct researches in various statistical aspects (Basseville and Nikiforov[1]).

In this chapter, using the concept of the sequential analysis approach, we develop an innovative Bayesian method designed specifically for the best solution in selection problem. The proposed method adopts the optimization concept of Bayesian inference and the uncertainty of the decision-making method in dynamic programming environment. The proposed algorithm is capable of taking into consideration the quality attributes of uncertain values in determining the optimal solution. Some authors have applied sequential analysis inference in combination with optimal stopping problem to maximize the probability of making correct decision. One of these researches is a new approach in probability distribution fitting of a given statistical data that Eshragh and Modarres [2] named it Decision on Belief (DOB). In this decision-making method, a sequential analysis approach is employed to find the best underlying probability distribution of the observed data. Moreover, Monfared and Ranaeifar [3] and Eshragh and Niaki [4] applied the DOB concept as a decision-making tool in some problems.

Since the idea behind the sequential analysis modeling is completely similar to the decision-making process of a human being in his life, it may perform better than available methods in decision-making problems. In these problems, when we want to make a decision, first we divide all of the probable solution space into smaller subspaces (the solution is one of the subspaces). Then based on our experiences, we assign a probability measure (belief) to each subspace, and finally we update the beliefs and make the decision.

## 2. AN APPLICATION TO DETERMINE THE BEST BINOMIAL DISTRIBUTION

In the best population selection problem, a similar decision-making process exits. First, the decision space can be divided into several subspaces (one for each population); second, the solution of the problem is one of the subspaces (the best population). Finally, we can assign a belief to each subspace where the belief denotes the performance of the population in term of its parameter. Based upon the updated beliefs in iterations of the data gathering process, we may decide which population possesses the best parameter value.

Consider $n$ independent populations $P_1, P_2, ..., P_n$, where for each index, population $i = 1, 2, ..., n$, is characterized by the value of its parameter of interest $P_i$. Let $p_{[1]} \leq ... \leq p_{[n]}$ denote the ordered value of the parameters $p_1, ..., p_n$. If we assume that the exact pairing between the ordered and the unordered parameter is unknown, then, a population $P_i$ with $p_i = p_{[n]}$ is called the best population.

There are many applications for the best population selection problem. As one application in supply chain environments, one needs to select the supplier among candidates that performs the best in terms of the quality of its products. As another example, in statistical analysis, we need to select a distribution among candidates that fits the collected observations the most. Selecting a production process that is in out-of-control state, selecting the stochastically optimum point of a multi-response problem, etc. are just a few of these applications.

The problem of selecting the best population was studied in papers by Bechhofer and Kulkarni [5] using the indifference zone approach and by Gupta and Panchapakesan [6] employing the best subset selection approach.

## 2.1. Belief and the approach of its improvement

Assume that there are $n$ available *Binomial* populations and we intend to select the one with the highest probability of success. Furthermore, in each stage of the data gathering process and for each population, we take an independent sample of size $m$. Let us define

$\alpha'_{i,t}$ and $\beta'_{i,t}$ to be the observed number of successes and failures of the $i^{th}$ *Binomial* population in the $t^{th}$ stage (sample) and

$\alpha_{i,k}$ and $\beta_{i,k}$ to be the cumulative observed number of successes and failures of the $i^{th}$ *Binomial* population up to the $k^{th}$ stage (sample) respectively. In other words,

$\alpha_{i,k} = \sum_{t=1}^{k} \alpha'_{i,t}$ and $\beta_{i,k} = \sum_{t=1}^{k} \beta'_{i,t}$. Then, in the $k^{th}$ stage defining $\overline{p_{i,k}}$ to be the estimated probability of success of the $i^{th}$ population obtained by $\frac{\alpha_{i,k}}{km}$, referring to Jeffrey's prior (Nair et al.[7]), for $\overline{p_{i,k}}$, we take a *Beta* prior distribution with parameters $\alpha_{i,0} = 0.5$ and $\beta_{i,0} = 0.5$. Then, using Bayesian inference, we can easily show that the posterior probability density function of $\overline{p_{i,k}}$ is

$$f(\overline{p_{i,k}}) = \frac{\Gamma(\alpha_{i,k} + \beta_{i,k} + 1)}{\Gamma(\alpha_{i,k} + 0.5)\Gamma(\beta_{i,k} + 0.5)} \overline{p_{i,k}}^{\alpha_{i,k} - 0.5} (1 - \overline{p_{i,k}})^{\beta_{i,k} - 0.5} \qquad (1)$$

At stage *k* of the data gathering process, after taking a sample and observing the numbers of failures and successes, we update the probability distribution function of $\overline{p_{i,k}}$ for each population. To do this, define $B(\alpha_{i,k}, \beta_{i,k})$ as a probability measure (called belief) of the $i^{th}$ population to be the best one given $\alpha_{i,k}$ and $\beta_{i,k}$ as

$$B(\alpha_{i,k}, \beta_{i,k}) = \Pr\{i^{th} \text{population is the best} | \alpha_{i,k}, \beta_{i,k}\} \qquad (2)$$

We then update the beliefs based on the values of $(\alpha_{i,k}, \beta_{i,k})$ for each population in iteration *k*. If we define $B(\alpha_{i,k-1}, \beta_{i,k-1})$ as the prior belief for each population, in order to update the posterior belief $B(\alpha_{i,k}, \beta_{i,k})$, since we may assume that the data are taken independently in each stage, we will have

$$B(\alpha_{i,k}, \beta_{i,k}) =$$

$$\frac{\Pr\{i^{th} \text{ Population is the best} | (\alpha_{i,k-1}, \beta_{i,k-1})\} \Pr\{(\alpha_{i,k}, \beta_{i,k}) | i^{th} \text{ Population is the best}\}}{\sum_{j=1}^{n}\left[\Pr\{j^{th} \text{ Population is the best} | (\alpha_{j,k-1}, \beta_{j,k-1})\} \Pr\{(\alpha_{j,k}, \beta_{j,k}) | j^{th} \text{ Population is the best}\}\right]}$$

$$= \frac{B(\alpha_{i,k-1}, \beta_{i,k-1}) \Pr\{(\alpha_{i,k}, \beta_{i,k}) | i^{th} \text{ Population is the best}\}}{\sum_{j=1}^{n}\left[B(\alpha_{j,k-1}, \beta_{j,k-1}) \Pr\{(\alpha_{j,k}, \beta_{j,k}) | j^{th} \text{ Population is the best}\}\right]} \qquad (3)$$

From equation (3) we see that to update the beliefs, we need to evaluate

$\Pr\{(\alpha_{i,k}, \beta_{i,k}) | i^{th}$ Population is the best$\}$ $i = 1, 2, ..., n$ in each decision-making stage. One way to do this is to use

$$\Pr\{(\alpha_{i,k}, \beta_{i,k}) | i^{th} \text{ Population is the best}\} = \frac{\overline{p_{i,k}}}{\sum_{j=1}^{n} \overline{p_{j,k}}} \qquad (4)$$

Then, the probability given in equation (3) will increase when a better population is selected. In the next theorem, we will prove that when the number of decision-making stages goes to infinity this probability converges to one for the best population.

**Theorem 1**
If the $i^{th}$ population is the best, then $\text{Lim } B(\alpha_{i,k}, \beta_{i,k}) = B_i = 1$.
In order to prove the theorem first we prove the following two lemmas.

**Lemma 1:**
Define a recursive sequence $\{R_{k,j}; j = 1, 2, ..., l\}$ as

$$R_{k,j} = \begin{cases} \dfrac{c_j R_{k-1,j}}{\sum_{i=1}^{l} c_i R_{k-1,i}} & \text{for } k = 1, 2, 3, ... \\ P_j & \text{for } k = 0 \end{cases} \qquad (5)$$

where
$c_1, c_2, ..., $ and $C_l$ are different positive constants,
$\sum_{j=1}^{l} P_j = 1$, and $P_j > 0$ Then, if

$l_j = \underset{k \to \infty}{Lim}(R_{k,j})$, there exist at most one non-zero $l_j$.

**Proof:**
Suppose there are two nonzero
$l_s > 0$ and $l_t > 0$. Taking the limit on $R_{k,j}$ as $k$ goes to infinity we have

$$\underset{k \to \infty}{Lim}(R_{k,j}) = l_j = \underset{k \to \infty}{Lim}\left(\dfrac{c_j R_{k-1,j}}{\sum_{i=1}^{l} c_i R_{k-1,i}}\right) = \dfrac{c_j l_j}{\sum_{i=1}^{l} c_i l_i} \qquad (6)$$

Now since
$l_s > 0$ and $l_t > 0$, then by equation (6) we have

$$l_s = \frac{c_s l_s}{\sum_{i=1}^{l} c_i l_i} \Rightarrow c_s = \sum_{i=1}^{l} c_i l_i \text{ and } l_t = \frac{c_t l_t}{\sum_{i=1}^{l} c_i l_i} \Rightarrow c_t = \sum_{i=1}^{l} c_i l_i \tag{7}$$

In other words, we conclude $c_s = c_t$, which is a contradiction.

**Lemma 2:**
Sequence $R_{k,j}$ converges to one for $j = g$ and converges to zero for $j \neq g$, where $g$ is an index for the maximum value of $c_j$.

**Proof**
From equation (6), we know that $\sum_{j=1}^{l} l_j = 1$. Then by lemma 1, we have $l_i = 1$ for only one $i$. Now suppose that $c_g = \max_{j \in \{1...m\}} \{c_j\}$ and $g \neq i$.. We will show that this is a contradiction.

By equation (5), we have $H_{k,i} = \frac{c_g}{c_i} H_{k-1,i}$. Since $\frac{c_g}{c_i} > 0$ we will have

$$H_{k,i} = \frac{c_g}{c_i} H_{k-1,i} = \left(\frac{c_g}{c_i}\right)^k H_{0,i} \Rightarrow \lim_{k \to \infty}(H_{k,i}) = \infty \tag{8}$$

That is a contradiction because $\lim_{k \to \infty}(H_{k,i}) = \frac{\lim_{k \to \infty}(R_{k,g})}{\lim_{k \to \infty}(R_{k,i})} = \frac{l_g}{l_i} = 0$. So $l_g = 1$

Now we are ready to prove the convergence property of the proposed method. Taking limit on both sides of equation (3), we will have

$$\lim_{k \to \infty} B(\alpha_{i,k}, \beta_{i,k}) = B_i = \lim_{k \to \infty} \left[ \frac{B(\alpha_{i,k-1}, \beta_{i,k-1}) \Pr\{(\alpha_{i,k}, \beta_{i,k}) | i^{th} \text{ Population is the best}\}}{\sum_{j=1}^{n} \left[ B(\alpha_{j,k-1}, \beta_{j,k-1}) \Pr\{(\alpha_{j,k}, \beta_{j,k}) | j^{th} \text{ Population is the best}\}\right]} \right]$$

$$\tag{9}$$

From the law of large numbers, we know that $\lim_{k \to \infty} \overline{p_{j,k}} = p_j$, where $p_j$ is the probability of success of the $j^{th}$ population. Hence, using equation (7) we have $B_i = \dfrac{B_i p_i}{\sum_{j=1}^{n} B_j p_j}$. Then assuming population $i$ is the best, i.e., it possesses the largest value of $p_j$'s, by lemma 1 and 2 we conclude that $B_i = 1$ and $B_j = 0$, $j \neq i$. This concludes the convergence property of the proposed method.

In real-world applications, since there is a cost associated with the data gathering process we need to select the best population in a finite number of decision-making stages. In the next section, we present the proposed decision-making method in the form of a stochastic dynamic programming model in which there is a limited number of decision-making stages available to select the best population.

## 2.2. A Dynamic Programming Approach

The proposed dynamic programming approach to model the decision-making problem of selecting the best Binomial population is similar to an optimal stopping problem.

Let us assume that to find the best population there is a limited number of stages (s) available. Then, the general framework of the decision-making process in each stage is proposed as:

Take an independent sample of size m from each population.

Calculate the posterior beliefs in terms of the prior beliefs using Bayesian approach.

Select the two biggest beliefs.

Based upon the values of the two biggest beliefs calculate the minimum acceptable belief.

If the maximum belief is more than the minimum acceptable belief, then we can conclude that the corresponding subspace is the optimal one. Otherwise, go to step 1.

In step 3 of the above framework, let populations $i$ and $j$ be the two candidates of being the best populations (it means that the beliefs of populations $i$ and $j$ are the two biggest beliefs) and we have $s$ decision-making stages. If the biggest belief is more than a threshold (minimum acceptable belief $d_{i,j}(s)$, $(0 \leq d_{i,j}(s) \leq 1)$, we select the corresponding subspace of that belief as the solution. Otherwise, the decision-making process continues by taking more observations. We determine the value of $d_{i,j}(s)$ such that the belief of making the correct decision is maximized. To do this suppose that for each population a new observation, $(\alpha_{j,k}, \beta_{j,k})$, is available at a given stage $k$. At this stage, we define $V(s, d_{i,j}(s))$ to be the expected belief of making the correct decision in $s$ stages when two populations $i$ and $j$ are the candidates for the optimal population. In other words, if we let $CS$ denote the event of making the correct decision, we define $V_{i,j}(s, d_{i,j}(s)) = E[B_{i,j}\{CS\}]$, where $B_{i,j}\{CS\}$ is the belief of making the correct decision. Furthermore, assume that the maximum of $V_{i,j}(s, d_{i,j}(s))$ occurs at $d_{i,j}^*(s)$. Then, we will have

$$V_{i,j}\left(s, d_{i,j}^*(s)\right) = \underset{d_{i,j}(s)}{Max}\left\{V_{i,j}\left(s, d_{i,j}(s)\right)\right\} = Max\left\{E\left[B_{i,j}\{CS\}\right]\right\} \quad (10)$$

We denote this optimal point by $V_{i,j}^*(s)$. In other words, $V_{i,j}^*(s) = V_{i,j}(s, d_{i,j}^*(s))$. Moreover, let us define $S_i$ and $S_j$ to be the state of selecting population $i$ and $j$ as the candidates for the optimal population,

$$V_{i,j}^*(s) = Max\left\{E\left[B_{i,j}\{CS\}\right]\right\} = \\ Max\left\{E\left[B_{i,j}\{CS|S_i\}B_{i,j}\{S_i\} + B_{i,j}\{CS|S_j\}B_{i,j}\{S_j\} + B_{i,j}\{CS|NS_{i,j}\}B_{i,j}\{NS_{i,j}\}\right]\right\}$$

$$(11)$$

In order to evaluate $V_{i,j}^*(s)$, in what follows we will find the belief terms of equation (11).

$B_{i,j}\{CS \mid S_i\}$ and $B_{i,j}\{CS \mid S_j\}$

These are the beliefs of making the correct decision if population $i$ or $j$ is selected as the optimal population, respectively. To make the evaluation easier, we denote these beliefs by $B_{i,j}(i)$ and $B_{i,j}(j)$. Then, using equation (2) we have

$$B_{i,j}\{CS|S_i\} = B_{i,j}(i) = \frac{B(\alpha_{i,k-1}, \beta_{i,k-1})\overline{p_{k,i}}}{B(\alpha_{i,k-1}, \beta_{i,k-1})\overline{p_{k,i}} + B(\alpha_{j,k-1}, \beta_{j,k-1})\overline{p_{k,j}}}$$

Similarly,

(12)

$$B_{i,j}\{CS|S_j\} = B_{i,j}(j) = \frac{B(\alpha_{j,k-1}, \beta_{j,k-1})\overline{p_{k,j}}}{B(\alpha_{j,k-1}, \beta_{j,k-1})\overline{p_{k,j}} + B(\alpha_{i,k-1}, \beta_{i,k-1})\overline{p_{k,i}}}$$

(13)

$B_{i,j}\{S_i\}$ and $B_{i,j}\{S_j\}$

These are the beliefs of selecting population $i$ or $j$ as the optimal population, respectively. Regarding the decision-making strategy, we have:

$$B_{i,j}(i) = \max(B_{i,j}(i), B_{i,j}(j)) \text{ and } B_{i,j}(i) \geq d_{i,j}^*(s) \qquad (14)$$

Hence, we define event $S_i$ as

$$S_i \equiv \{B_{i,j}(i) = \max\{B_{i,j}(i), B_{i,j}(j)\}, B_{i,j}(i) \geq d_{i,j}^*(s)\} \qquad (15)$$

Since $B_{i,j}(i) + B_{i,j}(j) = 1$ and that the beliefs are not negative we conclude max $\{B_{i,j}(i), B_{i,j}(j)\} \geq 0.5$. Furthermore, since the decision making is performed based upon the maximum value of the beliefs, without interruption of assumptions, we can change the variation interval of $d_{i,j}^*(s)$ from [0,1] to [0.5,1]. Now by considering $d_{i,j}^*(s) \geq 0.5$ implicitly, we have $S_i \equiv \{B_{i,j}(i) \geq d_{i,j}^*(s)\}$. By similar reasoning $S_j \equiv \{B_{i,j}(j) \geq d_{i,j}^*(s)\}$. Hence

$$B_{i,j}\{S_j\} = \Pr\{B_{i,j}(j) \geq d_{i,j}^*(s)\} =$$

$$\Pr\left\{\frac{B(\alpha_{j,k-1},\beta_{j,k-1})\overline{p_{j,k}}}{B(\alpha_{i,k-1},\beta_{i,k-1})\overline{p_{i,k}} + B(\alpha_{j,k-1},\beta_{j,k-1})\overline{p_{j,k}}} \geq d_{i,j}^*(s)\right\} = \Pr\left\{\overline{p_{j,k}} \geq h(d_{i,j}^*(s))\overline{p_{i,k}}\right\}$$

$$= \Pr\left\{\frac{\overline{p_{j,k}}}{\overline{p_{i,k}}} \geq h(d_{i,j}^*(s))\right\} \qquad (16)$$

In which, $h(d_{i,j}^*(s)) = \dfrac{d_{i,j}^*(s)B(\alpha_{i,k-1},\beta_{i,k-1})}{(1-d_{i,j}^*(s))B(\alpha_{j,k-1},\beta_{j,k-1})}$ To evaluate $\Pr\left\{\dfrac{\overline{p_{j,k}}}{\overline{p_{i,k}}} \geq h(d_{i,j}^*(s))\right\}$ in equation (16), let $f_1(\overline{p_{j,k}})$ and $f_2(\overline{p_{i,k}})$ to be the probability distributions of $\overline{p_{j,k}}$ and $\overline{p_{i,k}}$, respectively. Then,

$$f_2(\overline{p_{i,k}}) = \frac{\Gamma(\alpha_{i,k}+\beta_{i,k}+1)}{\Gamma(\alpha_{i,k}+0.5)\Gamma(\beta_{i,k}+0.5)}\overline{p_{i,k}}^{\alpha_{i,k}-0.5}(1-\overline{p_{i,k}})^{\beta_{i,k}-0.5}$$

$$f_1(\overline{p_{j,k}}) = \frac{\Gamma(\alpha_{j,k}+\beta_{j,k}+1)}{\Gamma(\alpha_{j,k}+0.5)\Gamma(\beta_{j,k}+0.5)}\overline{p_{j,k}}^{\alpha_{j,k}-0.5}(1-\overline{p_{j,k}})^{\beta_{j,k}-0.5} \qquad (17)$$

Hence,

$$\Pr\left\{\overline{p_{j,k}} \geq h(d_{i,j}^*(s))\overline{p_{i,k}}\right\} = \int_0^1\int_{h(d_{i,j}^*(s))\overline{p_{i,k}}}^1 f_1(\overline{p_{j,k}})f_2(\overline{p_{i,k}})d\overline{p_{j,k}}d\overline{p_{i,k}} =$$

$$\int_0^1\int_{h(d_{i,j}^*(s))\overline{p_{i,k}}}^1 A_i \overline{p_{i,k}}^{\alpha_{i,k}-0.5}(1-\overline{p_{i,k}})^{\beta_{i,k}-0.5} A_j \overline{p_{j,k}}^{\alpha_{j,k}-0.5}(1-\overline{p_{j,k}})^{\beta_{j,k}-0.5} d\overline{p_{j,k}}d\overline{p_{i,k}}$$

$$(18)$$

where

$$A_i = \frac{\Gamma(\alpha_{i,k}+\beta_{i,k}+1)}{\Gamma(\alpha_{i,k}+0.5)\Gamma(\beta_{i,k}+0.5)}, \quad A_j = \frac{\Gamma(\alpha_{j,k}+\beta_{j,k}+1)}{\Gamma(\alpha_{j,k}+0.5)\Gamma(\beta_{j,k}+0.5)}.$$

$$(19)$$

By change of variables technique, we have:

$$U = \frac{\overline{p_{j,k}}}{p_{i,k}} \text{ and } V = \overline{p_{i,k}}$$

$$f(U) = A_i A_j U^{\alpha_{i,k}-0.5} \int_0^1 V^{\alpha_{i,k}+\alpha_{j,k}} (1-V)^{\beta_{i,k}-1} (1-UV)^{\beta_{j,k}-0.5} dV$$

$$\Pr\left\{\frac{\overline{p_{j,k}}}{\overline{p_{i,k}}} \geq h(d_{i,j}^*(s))\right\} = 1 - \int_0^{h(d_{i,j}^*(s))} f(U) dU = 1 - F\left(h(d_{i,j}^*(s))\right)$$

(20)

For $B_{i,j}\{S_i\}$ we have

$$B_{i,j}\{S_i\} = \Pr\{B_{i,j}(i) \geq d_{i,j}^*(s)\} =$$
$$\Pr\{1 - B_{i,j}(j) \geq d_{i,j}^*(s)\} = \Pr\{B_{i,j}(j) \leq 1 - d_{i,j}^*(s)\} = F\left(h(1 - d_{i,j}^*(s))\right)$$

$B_{i,j}\{CS \mid NS_{i,j}\}$ is the belief of making the correct decision when none of the subspaces $i$ and $j$ has been chosen as the optimal one. In other words, the maximum beliefs has been less than $d_{i,j}^*(s)$ and the process of decision-making continues to the next stage. In terms of stochastic dynamic programming approach, the belief of this event is equal to the maximum belief of making the correct decision in $(s-1)$ stages. Since the value of this belief is discounted in the current stage, using discount factor α,

$$B_{i,j}\{CS \mid NS_{i,j}\} = \alpha V_{i,j}^*(s-1) \qquad (22)$$

Having all the belief terms of equation (11) evaluated in equations (12), (13), (14), (15), and (16), and knowing that by partitioning the state space we have $B_{i,j}\{NS_{i,j}\} = 1 - (B_{i,j}\{S_i\} + B_{i,j}\{S_j\})$, equation (11) can now be evaluated by substituting.

$$V_{i,j}^*(s) = \max_{0.5 \leq d_{i,j}(s) \leq 1} \{B_{i,j}(i)\Pr\{B_{i,j}(i) \geq d_{i,j}(s)\} + B_{i,j}(j)\Pr\{B_{i,j}(j) \geq d_{i,j}(s)\}$$
$$+ B_{i,j}\{NS_{i,j}|CS\}\big(1 - \Pr\{B_{i,j}(i) \geq d_{i,j}(s)\} - \Pr\{B_{i,j}(j) \geq d_{i,j}(s)\}\big)\}$$
$$= \max_{0.5 \leq d_{i,j}(s) \leq 1} \{B_{i,j}(i)\Pr\{B_{i,j}(i) \geq d_{i,j}(s)\} + B_{i,j}(j)\Pr\{B_{i,j}(j) \geq d_{i,j}(s)\} +$$
$$\alpha V_{i,j}^*(s-1)\big(1 - \Pr\{B_{i,j}(i) \geq d_{i,j}(s)\} - \Pr\{B_{i,j}(j) \geq d_{i,j}(s)\}\big)\}$$

(24)

### 2.2.1. Making The Decision

Assuming that for the two biggest beliefs we have $B_{i,j}(i) \geq B_{i,j}(j)$, equation (23) can be written as

$$V_{i,j}^*(s) = \big(B_{i,j}(i) - \alpha V_{i,j}^*(s-1)\big)\Pr\{B_{i,j}(i) \geq d_{i,j}^*(s)\} +$$
$$\big(B_{i,j}(j) - \alpha V_{i,j}^*(s-1)\big)\Pr\{B_{i,j}(j) \geq d_{i,j}^*(s)\} + \alpha V_{i,j}^*(s-1) \qquad (24)$$

For the decision-making problem at hand, three cases may happen

1) $B_{i,j}(i) < \alpha V_{i,j}^*(s-1)$

In this case, both $\big(B_{i,j}(i) - \alpha V_{i,j}^*(s-1)\big)$ and $\big(B_{i,j}(j) - \alpha V_{i,j}^*(s-1)\big)$ are negative. Since we are maximizing $V_{i,j}(s, d_{i,j}(s))$, then the two probability terms in equation (24) must be minimized. This only happens when we let $d_{i,j}^*(s) = 1$, making the probability terms equal to zero. Now since $B_{i,j}(i) < d_{i,j}^*(s) = 1$, we continue to the next stage.

2) $B_{i,j}(j) > \alpha V_{i,j}^*(s-1)$

In this case, $\big(B_{i,j}(i) - \alpha V_{i,j}^*(s-1)\big)$ and $\big(B_{i,j}(j) - \alpha V_{i,j}^*(s-1)\big)$ are both positive and to maximize $V_{i,j}(s, d_{i,j}(s))$ we need the two probability terms in equation (24) to be maximized. This only happens when we let $d_{i,j}^*(s) = 0.5$. Since $B_{i,j}(i) > d_{i,j}^*(s) = 0.5$

Since $B_{i,j}(i) > d_{i,j}^*(s) = 0.5$, we select population $i$ as the optimal subspace.

3) $B_{i,j}(j) > \alpha V_{i,j}^*(s-1)$:

In this case, one of the probability terms in equation (24) has positive coefficient and the other has negative coefficient. In this case, in order to maximize $V_{i,j}(s, d_{i,j}(s))$ we take the derivative as follows.

Substituting equations (20) and (21) in equation (24) we have

$$V_{i,j}(s, d_{i,j}(s)) = (B_{i,j}(i) - \alpha V_{i,j}^*(s-1))\{F(h(1 - d_{i,j}(s)))\} + (B_{i,j}(j) - \alpha V_{i,j}^*(s-1))\{1 - F(h(d_{i,j}(s)))\} + \alpha V_{i,j}^*(s-1) \quad (25)$$

Thus following is obtained,

$$V_{i,j}^*(s) = (B_{i,j}(i) - \alpha V_{i,j}^*(s-1))\Pr\left\{\overline{\frac{p_{j,k}}{p_{i,k}}} \le h(1 - d_{i,j}^*(s))\right\} + (B_{i,j}(j) - \alpha V_{i,j}^*(s-1))\Pr\left\{\overline{\frac{p_{j,k}}{p_{i,k}}} \ge h(d_{i,j}^*(s))\right\} + \alpha V_{i,j}^*(s-1) \quad (26)$$

For determining $\Pr\left\{\overline{\frac{p_{j,k}}{p_{i,k}}} \le h(1 - d_{i,j}^*(s))\right\}$, first using an approximation, we assume that $\overline{p_{i,k}}$ $\overline{p_{i,k}}$ is a constant number equal to its mean, then we have:

$$\Pr\left\{\frac{p_{j,k}}{p_{i,k}} \leq h\left(1-d_{i,j}^*(s)\right)\right\} =$$

$$\Pr\left\{\overline{p_{j,k}} \leq \overline{p_{i,k}}h\left(1-d_{i,j}^*(s)\right)\right\} = \int_0^{\overline{p_{i,k}}h\left(1-d_{i,j}^*(s)\right)} f_1\left(p_{j,k}\right)dp_{j,k} =$$

$$F_1\left(\overline{p_{i,k}}h\left(1-d_{i,j}^*(s)\right)\right)$$

$$\frac{\partial F_1\left(\overline{p_{i,k}}h\left(1-d_{i,j}^*(s)\right)\right)}{\partial d_{i,j}^*(s)} = \frac{B\left(\alpha_{i,k-1},\beta_{i,k-1}\right)}{\left(1-d_{i,j}^*(s)\right)^2 B\left(\alpha_{j,k-1},\beta_{j,k-1}\right)} f_1\left(\overline{p_{i,k}}h\left(1-d_{i,j}^*(s)\right)\right)$$

(27)

Also $\Pr\left\{\dfrac{p_{j,k}}{p_{i,k}} \geq h\left(d_{i,j}^*(s)\right)\right\}$ is obtained as follows,

$$\Pr\left\{\frac{\overline{p_{j,k}}}{p_{i,k}} \geq h\left(d_{i,j}^*(s)\right)\right\} = \Pr\left\{\overline{p_{j,k}} \geq \overline{p_{i,k}}h\left(d_{i,j}^*(s)\right)\right\} = \int_{\overline{p_{i,k}}h\left(d_{i,j}^*(s)\right)}^1 f_1\left(p_{j,k}\right)dp_{j,k} =$$

$$1 - F_1\left(\overline{p_{i,k}}h\left(d_{i,j}^*(s)\right)\right)$$

$$\frac{\partial\left(1 - F_1\left(\overline{p_{i,k}}h\left(d_{i,j}^*(s)\right)\right)\right)}{\partial d_{i,j}^*(s)} = \frac{-B\left(\alpha_{i,k-1},\beta_{i,k-1}\right)}{\left(d_{i,j}^*(s)\right)^2 B\left(\alpha_{j,k-1},\beta_{j,k-1}\right)} f_1\left(\overline{p_{i,k}}h\left(d_{i,j}^*(s)\right)\right)$$

(28)

Now following can be resulted,

$$\frac{\partial V_{i,j}(s,d_{i,j}(s))}{\partial d_{i,j}(s)}=0 \Rightarrow$$

$$(B_{i,j}(i)-\alpha V_{i,j}^*(s-1))\frac{B(\alpha_{i,k-1},\beta_{i,k-1})}{(1-d_{i,j}^*(s))^2 B(\alpha_{j,k-1},\beta_{j,k-1})_1} f_1\left(\overline{p_{i,k}}h(1-d_{i,j}^*(s))\right)=$$

$$-(B_{i,j}(j)-\alpha V_{i,j}^*(s-1))\frac{B(\alpha_{i,k-1},\beta_{i,k-1})}{(d_{i,j}^*(s))^2 B(\alpha_{j,k-1},\beta_{j,k-1})} f_1\left(\overline{p_{i,k}}h(d_{i,j}^*(s))\right) \Rightarrow$$

$$\frac{(B_{i,j}(i)-\alpha V_{i,j}^*(s-1))}{-(B_{i,j}(j)-\alpha V_{i,j}^*(s-1))}=\left(\frac{1-d_{i,j}^*(s)}{d_{i,j}^*(s)}\right)^{2(\alpha_{i,k-1}+\beta_{i,k-1})} \Rightarrow$$

$$d_{i,j}^1(s)=\frac{1}{\left(\frac{(B_{i,j}(i)-\alpha V_{i,j}^*(s-1))}{-(B_{i,j}(j)-\alpha V_{i,j}^*(s-1))}\right)^{\frac{1}{2(\alpha_{i,k-1}+\beta_{i,k-1})}}+1}$$

Now the approximate value of $d_{i,j}(s)$ say $d^1_{i,j}(s)$ is determined.

Second using another approximation, we assume that $\overline{p_{j,k}}$ is a constant number equal to its mean thus with similar reasoning, following is obtained:

$$d_{i,j}^2(s)=\frac{1}{\left(\frac{(B_{i,j}(j)-\alpha V_{i,j}^*(s-1))}{-(B_{i,j}(i)-\alpha V_{i,j}^*(s-1))}\right)^{\frac{1}{2(\alpha_{i,k-1}+\beta_{i,k-1})}}+1} \tag{30}$$

Therefore the approximate optimal value of $d_{i,j}^*(s)$ can be determined from following equation,

$$d_{i,j}^*(s)=Max\{d_{i,j}^1(s),d_{i,j}^2(s)\} \tag{31}$$

## 3. AN APPLICATION FOR FAULT DETECTION AND DIAGNOSIS IN MULTIVARIATE STATISTICAL QUALITY CONTROL ENVIRONMENTS

### 3.1. Introduction

In this section, a heuristic threshold policy is applied in phase II of a control charting procedure to not only detect the states of a multivariate quality control system, but also to diagnose the quality characteristic(s) responsible for an out-of-control signal. It is assumed that the in-control mean vector and in-control covariance matrix of the process have been obtained in phase I.

### 3.2. Background

In a multivariate quality control environment, suppose there are mm correlated quality characteristics whose means are being monitored simultaneously. Further, assume there is only one observation on the quality characteristics at each iteration of the data gathering process. where the goal is to detect the variable with the maximum mean shift. Let $x_{ki}$ be the observation of the $i^{th}$ quality characteristic, $i=1,2,...,mi=1,2,...,m$, at iteration $k$, $k=1,2,...$ and define the observation vector $x_k = [x_{k1}, x_{k2},..., x_{km}]^T$ and observation matrix $O_k = (x_1, x_2, ..., x_k)$. After taking a new observation, $x_k$, define $B_i(x_k, O_{k-1})$, the probability of variable ii to be in an out-of-control state, as

$$B_i(x_k, O_{k-1}) = \Pr\{OOC_i | x_k, O_{k-1}\}, \qquad (32)$$

where OOCOOC stands for out-of-control. This probability has been called the belief of variable ii to be in out-of-control condition given the observation matrix up to iteration k−1k−1 and the observation vector obtained at iteration kk.

Assuming the observations are taken independently at each iteration, to improve the belief of the process being in an out-of-control state at the $k^{th}$ iteration, based on the observation matrix $O_{k-1}$ and the new observation vector $x_k$, we have

$$\Pr\{x_k|OOC_i, O_{k-1}\} = \Pr\{x_k|OOC_i\} \qquad (33)$$

Then, using the Bayesian rule the posterior belief is:

$$B_i(x_k, O_{k-1}) = \Pr\{OOC_i|x_k, O_{k-1}\} = \frac{\Pr\{OOC_i, x_k, O_{k-1}\}}{\Pr\{x_k, O_{k-1}\}} =$$

$$\frac{\Pr\{OOC_i, x_k, O_{k-1}\}}{\sum_{j=1}^{m}\Pr\{OOC_j, x_k, O_{k-1}\}} = \frac{\Pr\{OOC_i|O_{k-1}\}\Pr\{x_k|OOC_i, O_{k-1}\}}{\sum_{j=1}^{m}\Pr\{OOC_j|O_{k-1}\}\Pr\{x_k|OOC_j, O_{k-1}\}} \qquad (34)$$

Since the goal is to detect the variable with the maximum mean shift, only one quality characteristic can be considered out-of-control at each iteration. In this way, there are $m-1$ remaining candidates for which $m-1$ quality characteristics are in-control. Hence, one can say that the candidates are mutually exclusive and collectively exhaustive. Therefore, using the Bayes' theorem, one can write equation (34) as

$$B_i(x_k, O_{k-1}) = \frac{\Pr\{OOC_i|O_{k-1}\}\Pr\{x_k|OOC_i\}}{\sum_{j=1}^{m}\Pr\{OOC_j|O_{k-1}\}\Pr\{x_k|OOC_j\}} = \frac{B_i(x_{k-1}, O_{k-2})\Pr\{x_k|OOC_i\}}{\sum_{j=1}^{m}B_j(x_{k-1}, O_{k-2})\Pr\{x_k|OOC_j\}}$$

$$(35)$$

When the system is in-control, we assume the $m$ characteristics follow a multinormal distribution with mean vector $\mu = [\mu_1, \mu_2, ..., \mu_m]^T$ and covariance matrix

$$\Sigma = \begin{bmatrix} \sigma_1^2 & \sigma_{12} & \cdot & \sigma_{1m} \\ \sigma_{21} & \sigma_2^2 & \cdot & \sigma_{2m} \\ \cdot & \cdot & \cdot & \cdot \\ \sigma_{m1} & \sigma_{m2} & \cdot & \sigma_m^2 \end{bmatrix} \quad (36)$$

In out-of-control situations, only the mean vector changes and the probability distribution along with the covariance matrix remain unchanged. In latter case, equation (35) is used to calculate the probability of shifts in the process mean

$\mu$ at different iterations. Moreover, in order to update the beliefs at iteration $k$ one needs to evaluate $\Pr\{x_k \mid OOC_i\}$.

The term $\Pr\{x_k \mid OOC_i\}$ is the probability of observing $x_k$ if only the $i^{th}$ quality characteristic is out-of-control. The exact value of this probability can be determined using the multivariate normal density, $A \exp\left(-\frac{1}{2}(x_k - \mu_{1i})^T \Sigma^{-1}(x_k - \mu_{1i})\right)$, where $\mu_{1i}$ denotes the mean vector in which only the $i^{th}$ characteristic has shifted to an out-of-control condition and $A$ is a known constant.

Since the exact value of the out-of-control mean vector $\mu_{1i}$ is not known a priori, two approximations are used in this research to determine $\Pr\{x_k \mid OOC_i\}$. Note that we do not want to determine the exact probability. Instead, the aim is to have an approximate probability (a belief) on each characteristic being out-of-control. In the first approximation method, define $IC_i$ to be the event that all characteristics are in-control, and let $\Pr\{x_k \mid IC_i\}$ be the conditional probability of observing $x_k$ given all characteristics are in-control. Further, $x'_k = [\mu_{01}, ..., x_{ki}, \mu_{0i+1}, ..., \mu_{0m}]^T$ in the aforementioned multivariate normal density, so that $\Pr\{x_k \mid IC_i\}$ can be approximately evaluated using $\Pr\{x_k \mid IC_i\} = \Pr\{x'_k \mid IC_i\}$, where $\Pr\{x'_k \mid IC_i\} = A \exp\left(-\frac{1}{2}(x'_k - \mu_0)^T \Sigma^{-1}(x'_k - \mu_0)\right)$. Note that this evaluation is proportional to $\exp\left(-\frac{1}{2}\left(\frac{x_{ki} - \mu_{0i}}{\sigma_i}\right)^2\right)$, and since it is assumed that characteristic $i$ is under control, no matter the condition of the other characteristics, this approximation is justifiable.

In the second approximation method, we assume $\Pr\{x_k \mid OOC_i\} \propto \frac{1}{\Pr\{x_k \mid IC_i\}}$. Although it is obvious that $\Pr\{x_k \mid OOC_i\}$ is not equal to $\frac{1}{\Pr\{x_k \mid IC_i\}}$, since we only need a belief function to evaluate $\Pr\{x_k \mid OOC_i\}$ and also we do not know the exact value of out-of-control mean vector, this approximation is just used to determine $\Pr\{x_k \mid OOC_i\}$. Moreover, it can be easily seen that the closer the value of the $i^{th}$ characteristic is to its in-control mean the smaller is $\Pr\{x_k \mid OOC_i\}$ as expected. We thus let

$$\Pr\{x_k \mid OOC_i\} \propto \frac{1}{\Pr\{x_k \mid IC_i\}} = R\exp\left(\frac{1}{2}\left(\frac{x_{ki}-\mu_{0i}}{\sigma_i}\right)^2\right) \; ; \; i=1,2,\ldots,m, \qquad (37)$$

where $R$ is a sufficiently big constant number to ensure the above definition is less than one.

The approximation to $\Pr\{x_k \mid OOC_i\}$ in equation (37) has the following two properties:

It does not require the value of out-of-control means to be known.

The determination of a threshold for the decision-making process (derived later) will be easier.

Niaki and Fallahnezhad [8] defined another equation for the above conditional probability and showed that if a shift occurs in the mean of variable ii, then $\lim_{k \to \infty} B_i(x_k, O_{k-1}) = B_i = 1$. They proposed a novel method of detection and classification and used simulation to compare its performances with that of existing methods in terms of the average run length for different mean shifts. The results of the simulation study were in favor of their proposed method in almost all shift scenarios. Besides using a different equation, the main difference between the current research and Niaki and Fallahnezhad [8] is that the current work develops a novel heuristic threshold policy, in which to save sampling cost and time or when these factors are constrained, the number of the data gathering stages is limited.

## 3.3. The Proposed Procedure

Assuming a limited number of the data gathering stages, $N$, to detect and diagnose characteristic(s), a heuristic threshold policy-based model is developed in this Section. The framework of the proposed decision-making process follows.

**Step I**

Define $i = 1, 2, ..., m$, as the set of indices for the characteristics, all of which having the potential of being out-of-control.

**Step II**

Using the maximum entropy principle, initialize $B_i(O_0) = 1/m$ as the prior belief of the $i^{th}$ variable to be out-of-control. In other words, at the start of the decision-making process all variables have an equal chance of being out-of-control. Set the discount rate $\alpha$, the maximum probability of correct selection when $N$ decision making stages remains $V(N)$, and the maximum number of decision making stages $N$.

**Step III**

Set $k = 0$

**Step IV**

Obtain an observation of the process.

**Step V**

Estimate the posterior beliefs, $B_i(O_k)$ (for $i = 1, 2, ..., m$), using equation (35).

**Step VI**

Obtain the order statistics on the posterior beliefs $B_i(O_k)$ such that

$$B_{(1)}(O_k) < B_{(2)}(O_k) < ... < B_{(m)}(O_k)$$

Furthermore, let $B_{gr}(O_k) = B_{(m)}(O_k)$ and $B_{sm}(O_k) = B_{(m-1)}(O_k)$.

**Step VII**

Assume the variables with the indices $i = gr$ and $j = sm$ are the candidates of being out-of-control, where $N$ decision-making steps are available.

Define $V(N, d_{i,j}(k))$ the probability of correct choice between the variables ii and jj, where $d_{i,j}(k)$ is the acceptable belief. Also, define $CS$ the event of correct selection and event $E_{i,j}$ the existence of two out-of-control candidate variables ii and jj. Then, we have:

$$V(N, d_{i,j}(k)) = \Pr\{CS|E_{i,j}\} \triangleq \Pr_{i,j}\{CS\} \tag{38}$$

where " $\triangleq$ " means "defined as."

Assuming $d^*_{i,j}(k)$ the maximum point of $V(N, d_{i,j}(k))$, called the minimum acceptable belief, we have

$$V(N, d^*_{i,j}(k)) \triangleq V^*_{i,j}(N) = \underset{d_{i,j}(k)}{Max}\{V(N, d_{i,j}(k))\} \triangleq Max\{\Pr_{i,j}\{CS\}\} \tag{39}$$

Let $S_i$ and $S_j$ be the event of selecting ii and jj as the out-of-control variables, respectively, and $NS_{i,j}$ be the event of not selecting any. Then, by conditioning on the probability, we have:

$$V^*_{i,j}(N) = Max\{\Pr_{i,j}\{CS\}\} =$$
$$Max\{\Pr_{i,j}\{CS|S_i\}\Pr_{i,j}\{S_i\} + \Pr_{i,j}\{CS|S_j\}\Pr_{i,j}\{S_j\} + \Pr_{i,j}\{CS|NS_{i,j}\}\Pr_{i,j}\{NS_{i,j}\}\} \tag{40}$$

At the $k^{th}$ iteration, the conditional bi-variate distribution of the sample means for variables grgr and $sm$, i.e, $X_{k, j=gr, sm} | X_{k, j \neq gr, sm}$, is determined using the conditional property of multivariate normal distribution given in appendix 1. Moreover, knowing $E(x_{k,j}) = \mu_j$ and evaluating the conditional mean and standard deviation (see appendix 1) results in

$$E(X_{k,i}|X_{k,j}) = \mu_i + \rho\frac{\sigma_i}{\sigma_j}(X_{k,j} - \mu_j) \tag{41}$$

and

$$E\left(X_{k,i}\big|X_{k,j}\right) = \mu_i + \rho\frac{\sigma_i}{\sigma_j}\left(X_{k,j} - \mu_j\right) \tag{42}$$

Based on the decomposition method of Mason et al. [9], define statistics $T_{k,j}$ and $T_{k,i|j}$ as

$$T_{k,j} = \left(\frac{X_{k,j} - \mu_j}{\sigma_j}\right) \tag{43}$$

$$T_{k,i|j} = \left(\frac{X_{k,i} - E\left(X_{k,i}\big|X_{k,j}\right)}{\sigma_{X_{k,i}|X_{k,j}}}\right) \tag{44}$$

Thus, when the process is in-control, the statistics $T_{k,j}$ and $T_{k,i|j}$ follow a standard normal distribution [9].

Now, let $B_{i,j}(i; x_k, O_{k-1})$ denote the probability of correct selection conditioned on selecting ii as the out-of-control variable. Hence,

$$B_{i,j}(i; x_k, O_{k-1}) = \frac{B_i(O_{k-1})e^{(0.5T_{k,i|j})^2}}{B_j(O_{k-1})e^{(0.5T_{k,j})^2} + B_i(O_{k-1})e^{(0.5T_{k,i|j})^2}}$$

$$B_{i,j}(j; x_k, O_{k-1}) = \frac{B_j(O_{k-1})e^{(0.5T_{k,j})^2}}{B_j(O_{k-1})e^{(0.5T_{k,j})^2} + B_i(O_{k-1})e^{(0.5T_{k,i|j})^2}} \tag{45}$$

Then, the probability measure $\Pr_{i,j}\{CS \mid S_i\}$ is calculated using the following equation,

$$\Pr_{i,j}\{CS|S_i\} = B_{i,j}(i;x_k,O_{k-1}) \qquad (46)$$

The probability measure $\Pr_{i,j}\{S_i\}$ is defined as the probability of selecting variable ii to be out-of-control. Regarding to the explained strategy, we have:

$$S_i \equiv \left\{ B_{i,j}(i;x_k,O_{k-1}) = Max\left\{ \begin{array}{c} B_{i,j}(i;x_k,O_{k-1}), \\ B_{i,j}(j;x_k,O_{k-1}) \end{array} \right\}, B_{i,j}(i;x_k,O_{k-1}) \geq d_{i,j}(k) \right\} \qquad (47)$$

Since $B_{i,j}(i;x_k, O_{k-1}) + B_{i,j}(j;x_k, O_{k-1}) = 1$ and the value of beliefs are not negative, we conclude

$$\max\{B_{i,j}(i;x_k,O_{k-1}), B_{i,j}(j;x_k,O_{k-1})\} \geq 0.5 \qquad (48)$$

Without interruption of assumptions, we can change the variation interval of $d_{i,j}(k)$ from [0,1] to [0.5,1]. Hence,

$$S_i = \{B_{i,j}(i;x_k,O_{k-1}) \geq d_{i,j}(k)\} \qquad (49)$$

By similar reasoning, we have:

$$S_j \equiv \{B_{i,j}(j;x_k,O_{k-1}) \geq d_{i,j}(k)\} \qquad (50)$$

The term $\Pr_{i,j}\{CS \mid NS_{i,j}\}$ denotes the probability of correct selection conditioned on excluding the candidates $i$ and $j$ as the solution. In other words, the maximum belief has been less than the threshold (minimum acceptable belief) $d_{i,j}^*(k)$ and the decision making process continues to the next stage. In terms of stochastic dynamic programming approach, the probability of this event is equal to the maximum probability of correct selection when there are $N-1$ stages remaining. The discounted value of this probability in the current stage using the discount factor $\alpha\alpha$ equals to $\alpha V_{i,i}(N-1)$. Further, since we partitioned the decision space into events $\{NS_{i,j}; S_i; S_j\}$, we have:

$$\Pr\nolimits_{i,j}\{NS_{i,j}\} = 1 - \left(\Pr\nolimits_{i,j}\{S_i\} + \Pr\nolimits_{i,j}\{S_j\}\right) \tag{51}$$

Now we evaluate $V_{i,j}^*(N)$ as follows,

$$V_{i,j}^*(N) =$$

$$\underset{0.5 \leq d_{i,j}(k) \leq 1}{\text{Max}} \left\{ \begin{array}{l} B_{i,j}(i;x_k,O_{k-1})\Pr\nolimits_{i,j}\{B_{i,j}(i;x_k,O_{k-1}) \geq d_{i,j}(k)\} + \\ B_{i,j}(j;x_k,O_{k-1})\Pr\nolimits_{i,j}\{B_{i,j}(j;x_k,O_{k-1}) \geq d_{i,j}(k)\} + \\ \Pr\nolimits_{i,j}\{NS_{i,j}|CS\} \left( \begin{array}{l} 1 - \Pr\nolimits_{i,j}\{B_{i,j}(i;x_k,O_{k-1}) \geq d_{i,j}(k)\} - \\ \Pr\{B_{i,j}(j;x_k,O_{k-1}) \geq d_{i,j}(k)\} \end{array} \right) \end{array} \right\}$$

$$= \underset{0.5 \leq d_{i,j}(k) \leq 1}{\text{Max}} \left\{ \begin{array}{l} B_{i,j}(i;x_k,O_{k-1})\Pr\nolimits_{i,j}\{B_{i,j}(i;x_k,O_{k-1}) \geq d_{i,j}(k)\} + \\ B_{i,j}(j;x_k,O_{k-1})\Pr\nolimits_{i,j}\{B_{i,j}(j;x_k,O_{k-1}) \geq d_{i,j}(k)\} + \\ \alpha V_{i,j}^*(N-1) \left( \begin{array}{l} 1 - \Pr\nolimits_{i,j}\{B_{i,j}(i;x_k,O_{k-1}) \geq d_{i,j}(k)\} - \\ \Pr\{B_{i,j}(j;x_k,O_{k-1}) \geq d_{i,j}(k)\} \end{array} \right) \end{array} \right\}$$

$$\tag{52}$$

In other words,

$$V_{i,j}^*(N) = \underset{0.5 \le d_{i,j}(k) \le 1}{Max} \begin{cases} B_{i,j}(i;O_k)\Pr_{i,j}\{B_{i,j}(i;O_k) \ge d_{i,j}(k)\} + \\ B_{i,j}(j;O_k)\Pr_{i,j}\{B_{i,j}(j;O_k) \ge d_{i,j}(k)\} + \\ \alpha V_{i,j}^*(N-1)\begin{pmatrix} 1 - \Pr_{i,j}\{B_{i,j}(i;O_k) \ge d_{i,j}(k)\} \\ -\Pr\{B_{i,j}(j;O_k) \ge d_{i,j}(k)\} \end{pmatrix} \end{cases}$$

The method of evaluating the minimum acceptable belief $d_{gr,sm}^*(k)$ is given in Appendix 2.

**Step VIII: The Decision Step**

If the belief $B_{gr,sm}(gr; x_k, O_{k-1})$ in the candidate set $(sm, gr)$) is equal to or greater than $d_{gr,sm}^*(k)$ then choose the variable with index grgr to be out-of-control. In this case, the decision-making process ends. Otherwise, without having any selection at this stage, obtain another observation, lower the number of remaining decision-stages to $N-1$, set $k = k+1$, and return to step **V** above. The process will continue until either the stopping condition is reached or the number of stages is finished. The optimal strategy with $N$ decision-making stages that maximizes the probability of correct selection would be resulted from this process.

In what follows, the procedure to evaluate $V_{i,j}^*(N)$ of equation (53) is given in detail.

## 3.4. Method of Evaluating $V_{i,j}^*(N)$

Using $d_{i,j}^*(k)$ as the minimum acceptable belief, from equation (53) we have

$$V_{i,j}^*(N) = \left(B_{i,j}(i;O_k) - \alpha V_{i,j}^*(N-1)\right)\Pr\left\{B_{i,j}(i;O_k) \geq d_{i,j}^*(k)\right\} +$$
$$\left(B_{i,j}(j;O_k) - \alpha V_{i,j}^*(N-1)\right)\Pr\left\{B_{i,j}(j;O_k) \geq d_{i,j}^*(k)\right\} + \alpha V_{i,j}^*(N-1) \quad (54)$$

## 5. CONCLUSION

In this chapter, we introduced a new approach to determine the best solution out of $m$ candidates. To do this, first, we defined the belief of selecting the best solution and explained how to model the problem by the Bayesian analysis approach. Second, we clarified the approach by which we improved the beliefs, and proved that it converges to detect the best solution. Next, we proposed a decision-making strategy using dynamic programming approach in which there were a limited number of decision-making stages.

## REFERENCES

1. Basseville, M. & Nikiforov, I.V. (1993). Detection of abrupt changes: Theory and application.(Information and System Sciences Series), Prentice-Hall. Using Dynamic Programming Based on Bayesian Inference in Selection Problems
2. Eshragh-J., A. & Modarres, M. (2009). A New Approach to Distribution Fitting: Decision on Beliefs, Journal of Industrial and Systems Engineering, 3(1), 56-71.
3. Saniee Monfared M.A., Ranaeifar F. (2007), Further Analysis and Developments of DOB Algorithm on Distribution Fitting; Scientia Iranica, 14(5), 425-434.
4. Eshragh-J., A. & Niaki, S.T.A. (2006). The application of decision on beliefs in response surface methodology. Proceedings of the 4th International Industrial Engineering Conference, Tehran, Iran.
5. Bechhofer, R.E. & Kulkarni, R. (1982). On the performance characteristics of closed adaptive sequential procedures for selecting the best *Bernoulli* populations. *Communication in Statistics – Sequential Analysis,* 1, 315-354.

6. Gupta, S.S. & Panchapakesan, S. (2002). Multiple decision procedures: Theory and methodology of selecting and ranking populations. Cambridge University Press, U.K.
7. Nair, V.N., Tang, B., and Xu, L. (2001). Bayesian inference for some mixture problems in quality and reliability, Journal of Quality Technology, 33, 16-28.
8. Niaki, S.T.A., Fallahnezhad, M. (2009), Decision-making in detecting and diagnosing faults in multivariate statistical quality control environments. International Journal of Advanced Manufacturing Technology 42, 713-724
9. Mason, R.L., Tracy, N.D., Young, J.C. (1995).Decomposition of T2 for multivariate control chart interpretation. *Journal of Quality Technology* 27, 99-108.

# CHAPTER 2

## INTELLIGENT CONDITION DIAGNOSIS METHOD BASED ON ADAPTIVE STATISTIC TEST FILTER AND DIAGNOSTIC BAYESIAN NETWORK

Ke Li [1], Qiuju Zhang [1,*], Kun Wang [1], Peng Chen [2] and Huaqing Wang [3]

[1]Jiangsu Key Laboratory of Advanced Food Manufacturing Equipment and Technology, Jiangnan University, 1800 Li Hu Avenue, Wuxi 214122, China

[2]Graduate School of Bioresources, Mie University/1577 Kurimamachiya-cho, Tsu, Mie 514-8507, Japan

[3]School of Mechanical & Electrical Engineering, Beijing University of Chemical Technology, Chao Yang District, Beijing 100029, China

## ABSTRACT

A new fault diagnosis method for rotating machinery based on adaptive statistic test filter (ASTF) and Diagnostic Bayesian Network (DBN) is presented in this paper. ASTF is proposed to obtain weak fault features under background noise, ASTF is based on statistic hypothesis testing in the frequency domain to evaluate similarity between reference signal (noise signal) and original signal, and remove the component of high similarity. The optimal level of significance α is obtained using particle swarm optimization (PSO). To evaluate the performance of the ASTF, evaluation factor $I_{pq}$ is also defined. In addition, a simulation experiment is designed to verify the effectiveness and robustness of ASTF. A sensitive evaluation method using principal component analysis (PCA) is proposed to evaluate the sensitiveness of symptom parameters (SPs) for condition diagnosis. By this way, the good SPs that have high sensitiveness for condition diagnosis can be selected. A three-layer DBN is developed to identify condition of rotation machinery based on the Bayesian Belief Network (BBN) theory. Condition diagnosis experiment for rolling element bearings demonstrates the effectiveness of the proposed method.

## KEYWORDS

feature extraction; adaptive statistic test filter; Diagnostic Bayesian Network; evaluation factor; condition diagnosis

## 1. INTRODUCTION

In the field of condition monitoring for rotating machinery, the vibration information, such as vibration accelerometer signal, vibration velocity signal, and vibration displacement signal, is often used for detecting faults and distinguishing fault types. Feature extraction of vibration signals is important for condition diagnosis [1,2]. However, feature extraction for condition diagnosis is difficult because the vibration signals measured for condition diagnosis contain strong noise component. Useful information is buried under stronger noise. In such case, the feature of machine condition could not be obtained and even the wrong conclusion will be induced. Thus, it is important that the feature of the signal can be sensitively extracted at the state change of a machine [3].

Many studies based on vibration signal processing technology have been carried out with the goal of machinery condition diagnosis [4,5,6,7]. Fourier analysis has been the dominating signal processing tool for condition diagnosis. In [8], Fourier analysis was used to identify the gear faults in planet cage. In [9], Fourier transform has been applied to detect rolling bearing faults. Unfortunately, there are some limitations of the Fourier transform, such as the fact that the signal to be analyzed must be strictly periodic or stationary. However, in practice, machinery operate under unsteady condition, such as varied rotating speed and operating load. In such case, even though the machinery is in the normal state, the spectrum feature and the frequency components of vibration signal are always changing with time. Thus the Fourier transform has no application to analyze non-stationary signal and can not reveal the inherent information of non-stationary signals [10,11]. Time frequency analysis methods, such as Wavelet Transforms (WT), Short Time Fourier Transform (STFT), *etc.*, are effective tools for analyzing the non-stationary signals. These technologies can simultaneously provide the joint distribution information of signals in time domain and frequency domain, and describe the energy density or intensity of the signal at different times and frequencies. In [12], STFT and Hilbert-Huang transform (HHT) analysis were integrated to detect faults of ball bearings for wind turbine. However, the result of the STFT method depends on the choice of the windows size. Moreover, computational cost the STFT is high. WT has got huge success in fault diagnostics of rotating machinery for its ability to focus on localized structures in time frequency domain. WT can decompose the signal into many basis functions and extract signal features through change of the scales and time shifts of the basis function. In [13], WT method was employed to extract fault features of external load changing and the asymmetry of three-phase in induction motor. In [14], the broken-bar fault of induction motor was detected based on discrete WT. However, the feature extraction results of WT rest with the choice of wavelet basis function. Only selecting the appropriate basis function, the features of signal can work well to detect faults. In addition, due to the limited length of the wavelet base function, energy loss is inevitable [15]. Empirical mode decomposition (EMD) technique was proposed by Huang*et al.*, for nonlinear and non-stationary signal processing. EMD is a self adaptive signal processing technology that could decompose a non-linear and non-stationary signal into a set of intrinsic mode functions (IMFs). However, undesired frequency components in results and undesired low amplitude IMFs at the low-frequency region remain unsolved in EMD [16,17,18,19]. In addition, there are many noise cancelling methods that have also been applied, such as band pass filter [20], Kalman filter [21], Wiener filter [22], and so on. However, due to their flaws and shortcomings, these methods cannot always be applied to failure feature extraction. For example, band pass filter cannot cancel the wide band noise; when using Wiener filter and Kalman filter to process signal, the signal must follow the normal distribution.

The number of the artificial intelligence techniques, such as artificial neural networks (ANN), ant colony optimization (ACO), Bayesian belief network (BBN) *etc.*, have been widely applied to fault diagnosis of plant machinery. In [23], three architecture NN, single-layer, multilayer perceptron network and counter propagation network, were introduced to detect

10 faults of a heat exchanger. In [2], an improved NN called partially-linearized neural network (PLNN) was presented to distinguishing the three types defect occurred in a rolling bearing. However, NN is not suitable for dealing with ambiguous diagnosis problems, and will never converge if SPs calculated by signals measured in different states have the same value. ACO algorithm imitates the behavior to solve optimization problems. In [24], ACO and DWT were integrated to detect faults of a rolling bearing used in the centrifugal fan system. However, ACO method is easy to trap into local optimum. In many cases, the optimization solution cannot be found. BBN is a powerful tool to represent and reason about complex systems with uncertain, incomplete and conflicting information [25,26]. A BBN enables us to model and reason about uncertainty, ideally suited for diagnosing real world problems where uncertain incomplete data exist. Therefore, it is a suitable solution for troubleshooting complex rotation machinery systems. In the last decades, BBN has been widely applied in condition diagnosis of plant machinery. References [27,28,29] are successful cases of BBN being used to detect fault of complex systems, such as nuclear power systems [27], aircraft engines [28], semiconductor manufacturing systems [29], *etc.*

To extract the fault feature of signals more effectively and discriminate conditions of rotation machinery more correctly, a novel method based on adaptive statistic test filter and Diagnostic Bayesian Network algorithm (DBN) for condition diagnosis of rotating machinery is presented. Structure of this paper is as follows: Section 2 instructs feature extraction method based on ASTF and evaluation factor $I_{pq}$. The optimal level of significance α is obtained by using PSO. In Section 3, the ten SPs for condition diagnosis are defined and PCA is employed to obtain high sensitive SPs for condition diagnosis. In Section 4, a three-layer DBN is built to identify condition of rotation machinery based on BBN theory. Section 5 shows a practical example of fault diagnosis for verifying the effectiveness of the proposed method. Summary and conclusions are given in Section 6.

## 2. FEATURE EXTRACTION BY ASTF

In this study, a new weak fault feature extraction method called adaptive statistic test filter (ASTF) is proposed. Principle of ASTF is based on statistic hypothesis testing in the frequency domain to evaluate similarity between reference signal (noise signal) and original signal, and remove the component of high similarity. Otherwise, the optimal level of significance

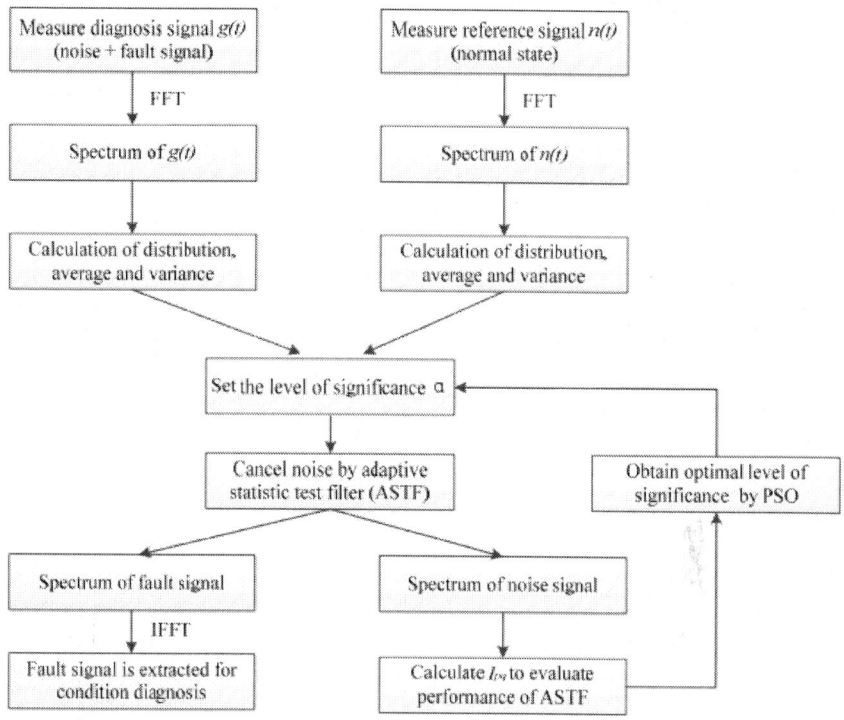

**Figure 1.** Procedure for applying the ASTF for the condition diagnosis

α is obtained using PSO. The procedure for applying STF for the condition diagnosis is proposed, as shown in Figure 1.

The reference signal $n(t)$ is measured in a normal state in advance. The original signal $g(t)$ is measured in the state to be detected and polluted by noise. $\mu_n(f)$ and $\mu_g(f)$ indicate the average value of $n(t)$ and $g(t)$ in the frequency domain, respectively; and $\sigma_n^2(f)$ and $\sigma_g^2(f)$ indicate the variance value of $n(t)$ and $g(t)$ in the frequency domain, respectively. The two null hypotheses are as follows:

$$H_1: \sigma^2{}_g(f) = \sigma^2{}_n(f) \tag{1}$$

$$H_0: \mu_g(f) = \mu_n(f) \tag{2}$$

Firstly, $\sigma^2{}_g(f) = \sigma^2{}_n(f)$ is verified by $F$ test with $m-1$ degree of freedom, and $F = S_n{}^2/S_g{}^2$.

$$S_g^2(f) = \sum_{j=1}^{m} g(f)^2/(m-1) - \overline{g}(f)^2 \qquad (3)$$

$$S_n^2(f) = \sum_{j=1}^{m} n(f)^2/(m-1) - \overline{n}(f)^2 \qquad (4)$$

If $\sigma^2{}_g(f) = \sigma^2{}_n(f)$, $\mu_g(f) = \mu_n(f)$ is verified by $t$ test with $2m-2$ degree of freedom, and $t = \{\overline{g}(f) - \overline{n}(f)\}/S\sqrt{2/m}$. here,

$$S^2 = \{(m-1)S_g^2 + (m-1)S_n^2\}/(2m-2) \qquad (5)$$

If $\sigma^2{}_g(f) \neq \sigma^2{}_n(f)$, $t = \{\overline{g}(f) - \overline{n}(f)\}/\sqrt{S_g^2(f)/m + S_n^2(f)/m}$.

$$m^* = \left(S_g^2/m + S_n^2/m\right)^2 / \left[\{S_g^4/m^2(m-1)\} + \{S_n^4/m^2(m-1)\}\right] \qquad (6)$$

If both null hypotheses would prove to be received, the spectrum component of the original signal $g(t)$ at the frequency $f$ is similar to that of the reference signal $n(t)$. The component at the frequency $f$ does not contain fault information and will be removed. If alternative hypothesis is denied, it means that the spectrum component of the original signal $g(t)$ at the frequency $f$ is not similar to that of the reference signal $n(t)$. The component at the frequency $f$ contains fault information.

After STF, the original signal $g(t)$ is decomposed into estimated fault signal $g^*(t)$ and estimated noise signal $n^*(t)$. In order to appraise the performance of STF, evaluation factor $I_{pq}$ is defined. $q_i$ and $q_i^*$ are the number that $n(t)$ and $n^*(t)$ cross over some level $i$ of the vertical coordinate of the power spectrum $F_n^2(f_k)$ with a positive slope in unit time and can be calculated as follows:

$$q_i = \frac{\sigma_v}{2\pi\sigma_x} e^{-n_i^2/2\sigma_x^2} \tag{7}$$

$$p_i = \frac{\sqrt{2\pi}}{\sigma_v} q_i \tag{8}$$

$$\sigma_x^2 = \int_0^\infty F_n^2(f_k)\, df_k \tag{9}$$

$$\sigma_v^2 = \int_0^\infty (2\pi f_k)^2 F_n^2(f_k)\, df_k \tag{10}$$

where i=1~K, ni=min{n(t)}~max{n(t)}.
$I_{pq}$ is defined as. follows:

$$I_{pq} = \sum_{i=1}^{K} \left|\log\left(\frac{q_i}{q_i^*}\right)\right|/K + \sum_{i=1}^{K} \left|\log\left(\frac{p_i}{p_i^*}\right)\right|/K \tag{11}$$

It is obvious that the smaller the value of the $I_{pq}$, the more similar $n(t)$ and $n^*(t)$ will be, and therefore, the better the STF will be. Thus, $I_{pq}$ is able to express the similarity degree between $n(t)$ and $n^*(t)$; that is to say, $I_{pq}$ can be used to evaluate the performance of STF.

To obtain optimal level of significance α, an adaptive PSO algorithm is proposed in this paper. PSO algorithm is based on groups, and solves an unconstrained D-dimensional optimization problem by minimization of the objective or the fitness function [30,31,32]. In this study, the fitness function is the evaluation factor $I_{pq}$ (Equation (11)). In PSO algorithm, each particle keeps track of its own position denoted by P(i).location=[Xi1,Xi2⋯XiD] and velocity denoted by P(i).velocity=[Vi1,Vi2⋯ViR] in the problem space, according to its own and neighboring particle experience [22,23,24]. The best previous position of particle is marked by the lowest fitness value and indicated by P(i).best=[Pi1,Pi2⋯PiR]. The best position among all particles experienced discovered by the swarm, so far, is defined as .best=[gi1,gi2⋯giR]. Then, the new positions and velocities of the particles are updated by the following equations:

$$P(i).velocity(t+1) = \omega\, P(i).velocity(t) + \eta_1 r_1[P(i).best(t) - P(i).location(t)] \\ + \eta_2 r_2[g(i)best(t) - P(i).location(t)] \quad (12)$$

$$P(i).location(t+1) = P(i).location(t) + P(i).velocity(t+1) \quad (13)$$

where $r_1$ and $r_2$ indicate random numbers between 0~1. $\eta_1$ is the cognitive parameter (acceleration coefficient). $\eta_2$ is the social parameter (acceleration coefficient). The inertia weight $\omega$ controls the previous velocity of particle, and $\omega$ adaptively adjust as follows:

$$\omega = \begin{cases} k_1 + 0.5q & R > 0.05 \\ k_2 + 0.5q & R \leqslant 0.05 \end{cases} \quad (14)$$

where $q$ is a random number with a uniform probability between 0~1; $k_1$ and $k_2$ are parameters, and $k_1 > k_2$, the choice of $k_1$ and $k_2$ is determined experimentally, here $k_1 = 0.5$ and $k_2 = 0.2$. R indicates change rate, which defined as Equation (15); if R is greater than 0.05, PSO is in the exploration stage, a large $\omega$ is beneficial to the algorithm's convergence; if R is less than 0.05, PSO is in the development stage, a small $\omega$ is beneficial to searching optimum point.

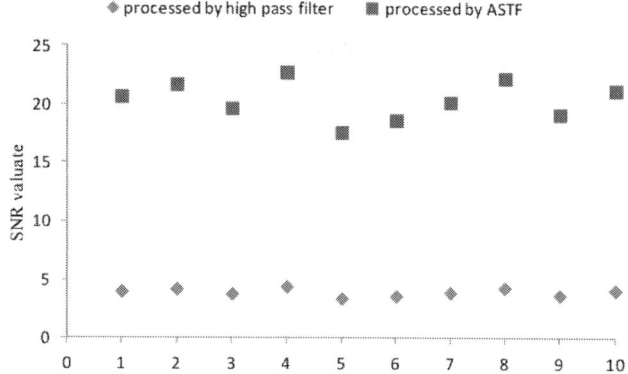

**Figure 2.** Signal-to-Noise Ratio (SNR) of denoised signals processed by each method

$$R = \frac{|I_{pq}(t+5) - I_{pq}(t)|}{|I_{pq}(t)|} \tag{15}$$

where $I_{pq}(t)$ is minimization evaluation factor value of the $t$-th iteration. $I_{pq}(t+5)$ is minimization evaluation factor value of the $(t+5)$-th iteration.

In order to test and verify capability of ASTF, a simulation experiment is designed. Ten set signals that consist of the impulsive signal with the period of 0.015 s and random white Gaussian noise are produced using Matlab software to simulate a bearing fault. These noisy signals are processed by ASTF and a high pass filter with 5000 Hz cut off frequency, respectively. The performances of denoising are estimated based on SNR. Mathematical expression of the impulsive signal is shown in Equation (16).

$$x(t) = x_0 e^{-\zeta \omega_n t} \sin \omega_n \sqrt{1 - \zeta^2} t \tag{16}$$

where $\zeta$ indicates coefficient of damping and $\zeta = 0.2$; $\omega_n$ expresses natural frequency and
$\omega_n = 3$ kHz; and $x_0$ denotes displacement constant and $x_0 = 2$.

Figure 2 shows the SNR of denoised signals processed by ASTF and high pass filter. As shown in Figure 2, all of the SNR values of denoised signals after ASTF are much greater than high pass filter. Then, ASTF method is effective and has high robustness for signal denoising.

## 3. SYMPTOM PARAMETERS FOR FAULT DIAGNOSIS AND SENSITIVITY EVALUATION

### 3.1. Symptom Parameters for Fault Diagnosis

The number of SPs reflect plant machinery condition have been defined in the pattern recognition field [33]. In this study, ten SPs in the time domain are considered.

$$P_1 = \frac{\sigma}{\bar{x}} \tag{17}$$

$$P_2 = \frac{\sum_{i=1}^{N} x_i^2}{\sigma^2} \tag{18}$$

$$P_3 = \frac{\left|\sum_{i=1}^{N} (x_i - \bar{x})^3\right|}{N\sigma^3} \tag{19}$$

$$P_4 = \frac{\sum_{i=1}^{N} (x_i - \bar{x})^4}{N\sigma^4} \tag{20}$$

where, $x_i$ is digital data of vibration signal. $\bar{x}$ is the mean value of $x_i$, $\bar{x} = \frac{\sum_{i=1}^{N} x_i}{N}$. $\sigma$ is standard deviation of $x_i$, $\sigma = \sqrt{\frac{\sum_{i=1}^{N} (x_i - \bar{x})^2}{N-1}}$.

$$P_5 = \frac{\overline{x_p}}{\overline{x}} \tag{21}$$

$$P_6 = \frac{\overline{x_p}}{\sigma} \tag{22}$$

$$P_7 = \frac{\left|\sum_{i=1}^{N_p}(x_{pi}-\overline{x_p})^3\right|}{N_p\sigma_p^3} \tag{23}$$

$$P_8 = \frac{\left|\sum_{i=1}^{N_p}(x_{pi}-\overline{x_p})^4\right|}{N_p\sigma_p^4} \tag{24}$$

where, $x_{pi}$ is the peak value of $x_i$. $\overline{x}_p$ and σp are the mean value and standard deviation of $x_{pi}$, respectively.

$$P_9 = \frac{\left|\sum_{i=1}^{N_v}(x_{vi}-\overline{x_v})^3\right|}{N_v\sigma_v^3} \tag{25}$$

$$P_{10} = \frac{\left|\sum_{i=1}^{N_v}(x_{vi}-\overline{x_v})^4\right|}{N_v\sigma_v^4} \tag{26}$$

where, $x_{vi}$ is the valley value of $x_i$. $\overline{x}_v$ and σv are the mean value and standard deviation of $x_{vi}$, respectively.

## 3.2. High Sensitivity Symptom Parameters Obtained by PCA

PCA is a statistical analytical tool used to explore, sort and group data. PCA takes a large number of correlated variables and transform these data into a smaller number of uncorrelated variables known as principal components. The first few principal components contain most of the information and the discriminatory features [34].

Define a data matrix with size $m \times n$, where $m$ is the number of identifying states and $n$ is the number of SPs, whose covariance matrix has eigenvalue $\lambda_i$ and eigenvector $a_i$ ($a$ is loading of the principal component and can express the importance of the SPs for each principal component) and $I = 1 - n$ with $\lambda_1 \geq \lambda_2 \geq ... \geq \lambda_n$. Principal components $Z_i$ and the cumulative contribution rate of the principal components $\eta_i$ can be calculated as follows:

$$\left\{ \begin{array}{c} Z_1 \\ \vdots \\ Z_n \end{array} \right\} = \left[ \begin{array}{ccc} a_{11} & \cdots & a_{1n} \\ \vdots & \ddots & \vdots \\ a_{m1} & \cdots & a_{mn} \end{array} \right] = AP \qquad (27)$$

$$\eta_i = \sum_{j=1}^{i} \lambda_j / \sum_{k=1}^{n} \lambda_k \qquad (28)$$

where $P_i$ indicates a symptom parameter, $I = 1 - n$.

## 4. BAYESIAN BELIEF NETWORK

BBN is a probability network based on graphical network model for describing causal uncertainties between variables. It is built for uncertainty modeling and reasoning, and has a great advantage in diagnosing fault caused by uncertainty and correlation of the complex systems.

## 4.1. Bayesian Inference

Supposing $A$ is a random event and $B$ is the event that is root causes generating $A$, conditional probabilities $P(A|B)$ between $A$ and $B$ can be calculated as follows:

$$P(A|B) = \frac{P(AB)}{P(B)} = \frac{P(A)P(B|A)}{P(B)} \tag{29}$$

where $P(AB)$ is the joint probability, $P(AB) = P(B) \cdot P(A|B) = P(A) \cdot P(B|A)$. Supposing $B_i$ ($i = 1, 2, ..., n$) are mutually exclusive and complete set of root causes generating $A$, the marginal probability of $A$ is

$$P(A) = \sum_{i=1}^{n} P(B_i) P(A|B_i) \tag{30}$$

The conditional and marginal probabilities of $A$ and $B_i$ is

$$P(B_i|A) = \frac{P(AB_i)}{P(A)} = \frac{P(B_i)P(A|B_i)}{\sum_{i=1}^{n} P(B_i)P(A|B_i)} \tag{31}$$

In Equation (31), $P(B_i)$ and $P(A/B_i)$ express prior probabilities and prior conditional probabilities, respectively. $P(B_i/A)$ indicates posterior probability. When using the Bayesian inference for fault diagnosis, $B_i$ represents an equipment condition and $A$ represents a SP. The prior probability of the SP $B_i$($P(B_i)$) and the conditional probability of equipment condition $A$ given $B_i P(A/B_i)$ can be obtained from expert experience or statistical data. Then, the posterior probability $P(B_i/A)$ can be obtained by Equation (31). If this posterior probability is high, the condition $B_i$ can be confirmed at the given $A$, and the equipment condition is judged $B_i$.

**Figure 3.** General model of Bayesian Belief Network.

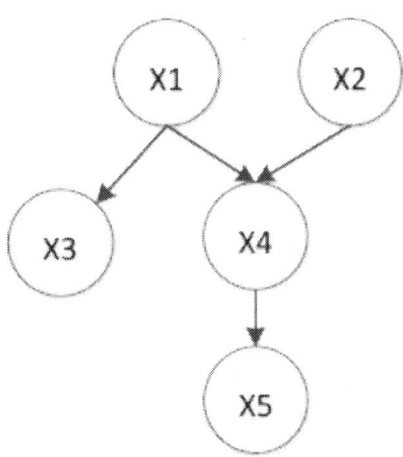

## 4.2. Topology of Bayesian Belief Network

A BBN consists of a number of nodes, directed links, and probability tables. For a diagnostic BBN model, nodes represent variables that can be SP, equipment condition or observations. Directed links indicate casual relationships between the variables. In this paper, the purpose of building a diagnostic BBN is to reason the most likely mechanical condition based on the values of SP, given one or more SP values to calculate posterior probabilities of the cause. The calculus of posterior probability involves calculating the joint probability for the model (probabilities of all combined states for all nodes within the model). The network contains five nodes, $X1$, $X2$, $X3$, $X4$, and $X5$, with a structure of three layers (see in Figure 3). In terms of the definition of the three types of conditional independence, $X1$ is independent of $X2$; $X1$ is parent of $X3$ and $X4$. Given $X1$, $X3$ and $X4$ are conditionally independent of each other, $X5$ is independent of $X1$, $X2$, and $X3$. The following derivation indicates how to calculate the posterior conditional probability $P(X4 = true|X5 = true)$.

$$P(X_4 = \text{true}| X_5 = \text{true}) = \frac{P(X_4 = \text{true}| X_5 = \text{true})}{P(X_5 = \text{true})}$$

$$= \frac{\sum_{x_1 x_2 x_3} P(X_1, X_2, X_3, X_4 = \text{true}, X_5 = \text{true})}{\sum_{x_1 x_2 x_3 x_4} P(X_1, X_2, X_3, X_4, X_5 = \text{true})} \tag{32}$$

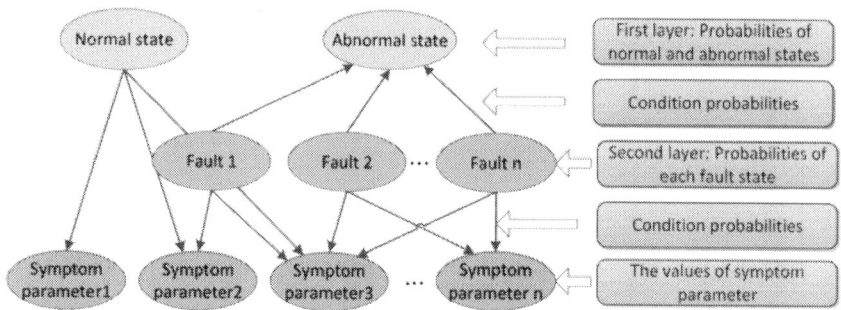

**Figure 4.** Structure of Diagnostic Bayesian Network.

where $P(X1, X2, X3, X4 = true|X5 = true)$ and $P(X1, X2, X3, X4, X5 = true)$ involve calculating the joint probability of the model. The joint probability of this model $P(X1, X2, X3, X4, X5)$ can be calculated as follows:

$$P(X_1, X_2, X_3, X_4, X_5) = \prod_{i=1}^{5} P(X_i| X_1, \cdots, X_{i-1})$$
$$= P(X_1)P(X_2| X_1)P(X_3| X_1 X_2)P(X_4| X_1 X_2 X_3)P(X_5| X_1 X_2 X_3 X_4) \quad (33)$$

Applying the independence assumption, the joint probability distribution can be simplified as follows:

$$P(X_1, X_2, X_3, X_4, X_5) = \prod_{i=1}^{5} P(X_i|P_{ai}) \quad (34)$$
$$= P(X_1)P(X_2)P(X_3| X_1)P(X_4| X_1 X_2)P(X_5| X_4)$$

## 4.3. Framework of Diagnostic Bayesian Network

In this study, Diagnostic Bayesian Network (DBN) is constructed for intelligent condition diagnosis. As shown in Figure 4, the proposed DBN consists of three layers. The first layer is normal and abnormal states. The sec-

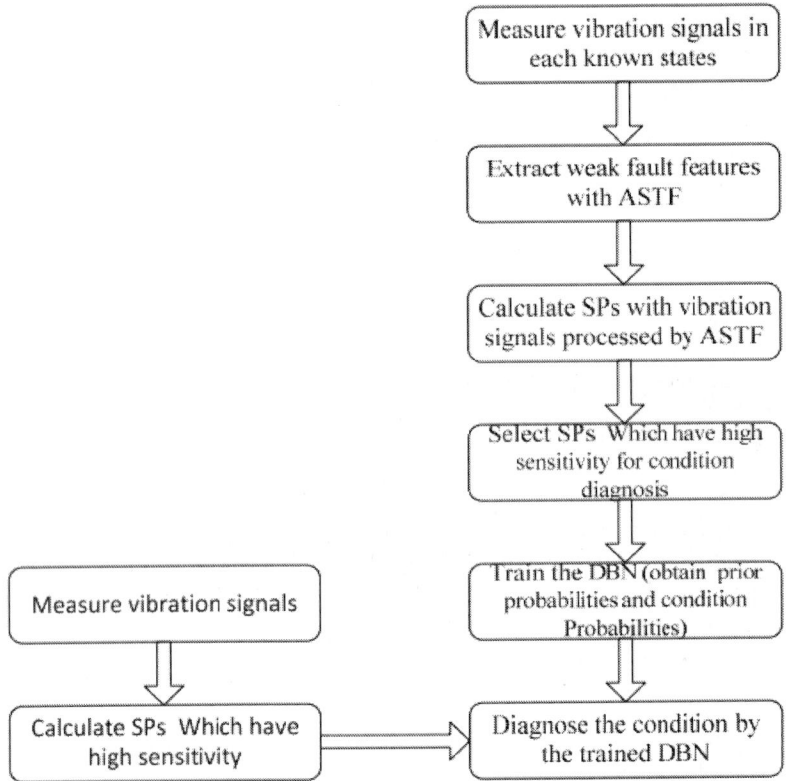

**Figure 5.** Flowchart for the condition diagnostic procedure.

ond layer is fault states, and the last layer is SPs calculated from the signals processed by ASTF.

In the proposed DBN, prior probabilities of root nodes are needed and conditional probabilities are also needed to represent direct probabilistic dependences among nodes in the three layers. Here, all machine states are regarded as parent nodes, and the prior probabilities of state $i$ ($S_i$) can be obtained as follows:

$$P(S_i) = \frac{N_{S_i}}{N} \tag{35}$$

where $N_{S_i}$ represents the sample size of state $i$, and $N$ indicates the total number of samples.

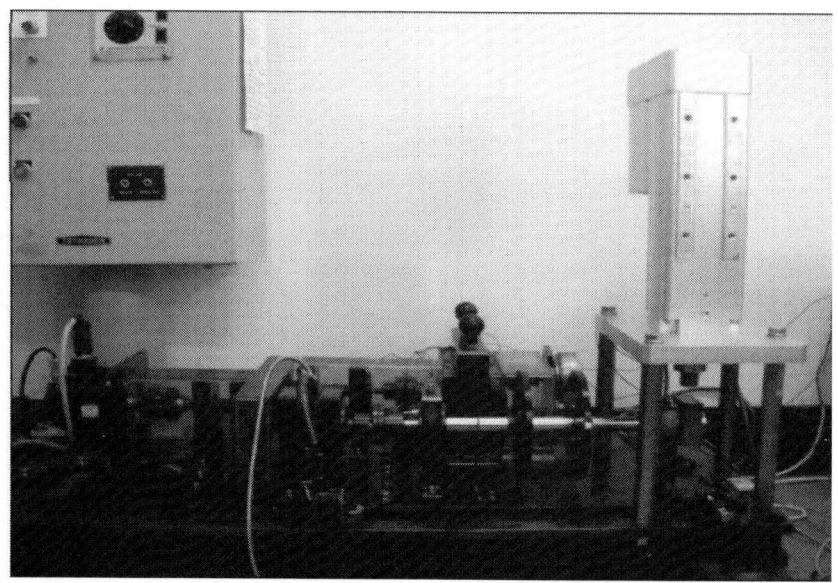

**Figure 6.** Experimental system for bearing fault diagnosis.

**Figure 7.** Bearing defects: (**a**) outer-race defect; (**b**) inner-race defect; and (**c**) roller defect.

The conditional probabilities of each node are obtained as follows

$$P(SP = x_i | S_i) = \frac{N_{S_i}^{x_i}}{N_{S_i}} \quad N_{S_i}^{x_i} \neq 0$$

$$P(SP = x_i | S_i) = \frac{1/N}{N_{S_i} + N_{x_i}/N} \quad if \quad N_{S_i}^{x_i} = 0 \tag{36}$$

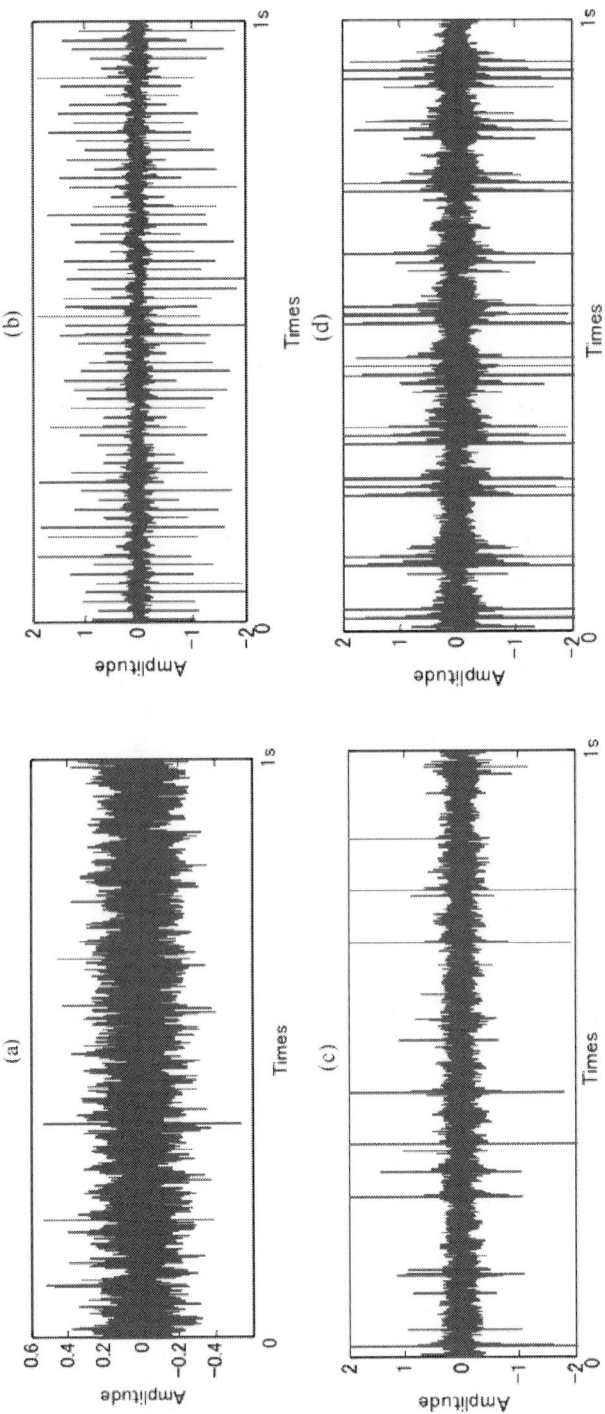

**Figure 8.** Original vibration signal: (**a**) normal state; (**b**) outer-race defect; (**c**) inner-race defect; and (**d**) roller defect.

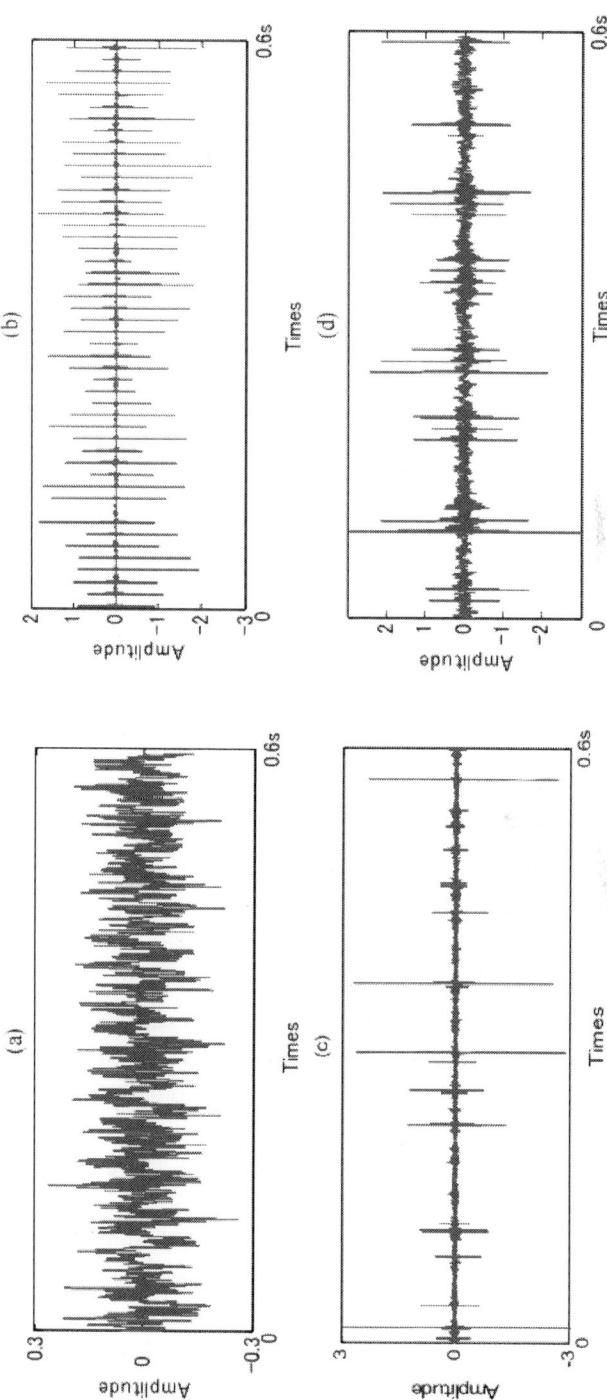

**Figure 9.** Vibration signal after ASTF: (**a**) normal state; (**b**) outer-race defect; (**c**) inner-race defect; and (**d**) roller defect.

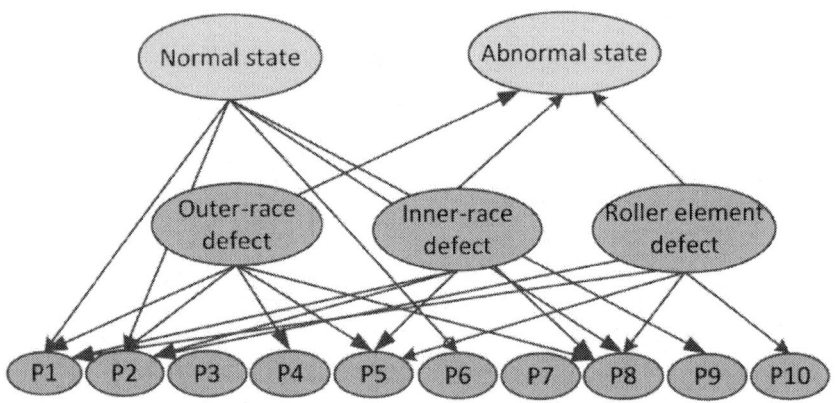

**Figure 10.** DBN for distinguishing conditions of a rolling bearing.

where *SP* represents the values of symptom parameters, and Nxi-Si indicates the sample size of state $i$, when $SP = x_i$.

## 5. DIAGNOSIS AND APPLICATION

### 5.1. Condition Diagnosis by Proposed Method

In this section, an experimental setup is designed to evaluate the effectiveness of the method proposed in this paper. The flowchart of the condition diagnostic procedure is shown in Figure 5.

Figure 6 shows the experimental bench for condition diagnosis test, which includes a servo motor, rotor system and loading equipment. The NSK 205 ball bearing is used for bearing condition diagnosis. As shown in Figure 7, three types fault: the outer defect, the inner defect, and the roller element defect were artificially made by using electro discharge machining with fault width was 0.3 mm, and fault depth was 0.025 mm.

In the present work, the original vibration signals in each state were measured by the accelerometer (PCB MA352A60, PCB Piezotronics Inc., New York, NY, USA) with 50,000 Hz sampling frequency. The accelerometer was fixed on vertical direction of the bearing. While the vibration signals were being obtained, the speed of servo motor was 800 rpm, and a 150 kg load was also transported on the rotating shaft by the loading equipment (RCS2-RA13R, IAI Co. Ltd., Shizuoka, Japan). All the data were recorded and transformed by a collection system includes a sensor signal conditioner

**Table 1.** First principal component of SPs.

| | Weight Coefficients for Each Symptom Parameter | | | | | | | | | | Contribution Rate | |
|---|---|---|---|---|---|---|---|---|---|---|---|---|
| | $P_1$ | $P_2$ | $P_3$ | $P_4$ | $P_5$ | $P_6$ | $P_7$ | $P_8$ | $P_9$ | $P_{10}$ | $\eta_1$ | |
| N:O | 0.99 | 0.91 | −0.62 | −0.75 | −0.33 | 0.85 | −0.69 | 0.88 | 0.79 | 0.21 | 0.88 |
| N:I | 0.93 | 0.87 | −0.53 | −0.32 | −0.65 | 0.78 | 0.11 | 0.92 | 0.81 | −0.42 | 0.86 |
| N:R | 0.99 | 1.0 | −0.9 | −0.36 | −0.75 | 0.9 | −0.22 | 0.86 | 0.91 | −0.36 | 0.89 |
| O:I | 0.87 | 1.0 | −0.56 | 0.99 | 0.88 | −0.66 | −0.71 | 0.98 | −0.56 | 0.38 | 0.90 |
| O:R | 0.99 | 0.99 | −0.37 | 0.93 | 0.95 | −0.58 | −0.62 | 0.98 | −0.79 | 0.86 | 0.85 |
| I:R | 0.97 | 0.86 | −0.78 | 0.36 | 0.91 | −0.55 | −0.11 | 0.87 | −0.91 | 0.95 | 0.86 |

**Table 2.** Selection result of the SPs for distinguishing each state.

| State | Selection result |
|---|---|
| Normal | $P_1, P_2, P_6, P_8, P_9$ |
| Outer-race defect | $P_1, P_2, P_4, P_5, P_8,$ |
| Inner-race defect | $P_1, P_2, P_5, P_8,$ |
| Roller-element defect | $P_1, P_2, P_5, P_8, P_{10},$ |

**Table 3.** Training sample data.

| State | Non-dimensional Symptom Parameters | | | | | | | | | |
|---|---|---|---|---|---|---|---|---|---|---|
| | $P_1$ | $P_2$ | $P_3$ | $P_4$ | $P_5$ | $P_6$ | $P_7$ | $P_8$ | $P_9$ | $P_{10}$ |
| Normal | 1 | 1 | 1 | 1 | 1 | 1 | 1 | 1 | 1 | 1 |
| | 1 | 1 | 1 | 1 | 1 | 2 | 1 | 1 | 1 | 2 |
| | 1 | 1 | 1 | 1 | 1 | 2 | 1 | 2 | 1 | 1 |
| | 1 | 1 | 1 | 1 | 1 | 1 | 1 | 1 | 1 | 2 |
| | ... | ... | ... | ... | ... | ... | ... | ... | ... | ... |
| Outer-race defect | 2 | 3 | 1 | 4 | 4 | 3 | 1 | 3 | 2 | 3 |
| | 3 | 3 | 1 | 4 | 4 | 3 | 1 | 3 | 2 | 4 |
| | 2 | 2 | 1 | 3 | 4 | 3 | 1 | 3 | 2 | 5 |
| | 3 | 3 | 1 | 4 | 4 | 3 | 1 | 3 | 2 | 3 |
| | ... | ... | ... | ... | ... | ... | ... | ... | ... | ... |
| Inner-race defect | 3 | 1 | 2 | 1 | 1 | 2 | 1 | 3 | 4 | 4 |
| | 4 | 1 | 2 | 2 | 1 | 5 | 5 | 4 | 5 | 4 |
| | 3 | 1 | 1 | 2 | 2 | 4 | 3 | 4 | 4 | 4 |
| | 3 | 1 | 1 | 1 | 1 | 3 | 3 | 3 | 3 | 2 |
| | ... | ... | ... | ... | ... | ... | ... | ... | ... | ... |
| Roller element defect | 5 | 5 | 1 | 3 | 5 | 5 | 3 | 4 | 4 | 5 |
| | 4 | 4 | 2 | 4 | 5 | 5 | 3 | 5 | 5 | 5 |
| | 4 | 4 | 5 | 3 | 5 | 5 | 3 | 4 | 3 | 5 |
| | 4 | 4 | 2 | 3 | 5 | 5 | 3 | 4 | 4 | 5 |
| | ... | ... | ... | ... | ... | ... | ... | ... | ... | ... |

**Table 4.** Diagnosis results of normal state.

| Non-Dimensional Symptom Parameters | | | | | State | | Judge |
|---|---|---|---|---|---|---|---|
| $P_1$ | $P_2$ | $P_6$ | $P_8$ | $P_9$ | Normal | Abnormal | |
| 1 | 1 | 1 | 1 | 1 | 0.96 | 0.04 | Normal |
| 1 | 1 | 1 | 2 | 1 | 0.89 | 0.11 | Normal |
| 1 | 1 | 2 | 1 | 1 | 0.91 | 0.09 | Normal |
| ... | ... | ... | ... | ... | ... | ... | ... |

**Table 5.** Diagnosis results of outer-race defect state.

| Non-Dimensional Symptom Parameters | | | | | State | | Judge |
|---|---|---|---|---|---|---|---|
| $P_1$ | $P_2$ | $P_4$ | $P_5$ | $P_8$ | Outer-Race Defect | Other Faults | |
| 3 | 3 | 4 | 4 | 3 | 0.99 | 0.01 | Outer-race defect |
| 2 | 2 | 4 | 4 | 3 | 0.89 | 0.11 | Outer-race defect |
| 2 | 3 | 5 | 4 | 3 | 0.86 | 0.14 | Outer-race defect |
| ... | ... | ... | ... | ... | ... | ... | ... |

**Table 6.** Diagnosis results of inner-race defect state.

| Non-dimensional Symptom Parameters | | | | State | | Judge |
|---|---|---|---|---|---|---|
| $P_1$ | $P_2$ | $P_5$ | $P_8$ | Inner-Race Defect | Other Faults | |
| 4 | 1 | 2 | 3 | 0.85 | 0.15 | Inner-race defect |
| 3 | 2 | 1 | 4 | 0.79 | 0.21 | Inner-race defect |
| 3 | 1 | 2 | 4 | 0.88 | 0.12 | Inner-race defect |
| ... | ... | ... | ... | ... | ... | ... |

**Table 7.** Diagnosis results of roller element defect state.

| Non-dimensional Symptom Parameters | | | | | State | | Judge |
|---|---|---|---|---|---|---|---|
| $P_1$ | $P_2$ | $P_5$ | $P_8$ | $P_{10}$ | Roller-Element Defect | Other Faults | |
| 4 | 4 | 5 | 5 | 5 | 0.75 | 0.25 | Roller-element defect |
| 3 | 3 | 4 | 5 | 5 | 0.68 | 0.32 | Roller-element defect |
| 3 | 4 | 5 | 4 | 5 | 0.66 | 0.34 | Roller-element defect |
| ... | ... | ... | ... | ... | ... | ... | ... |

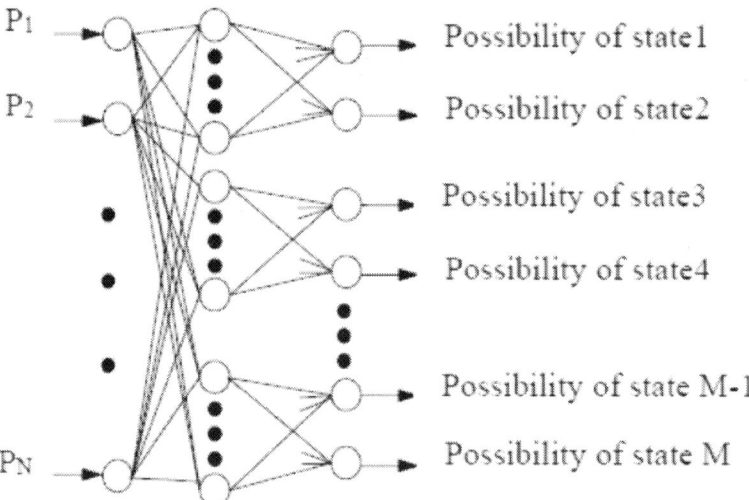

**Figure 11.** The back propagation NN for condition diagnosis.

(PCB ICP Model 480C02, PCB Piezotronics Inc., New York, NY, USA) and a signal recorder (Scope Coder DL750, YOKOGAWA Co. Ltd. Tokyo, Japan). Obtained data was divided to two sets, one set includes 80 samples and was used to train diagnosis system; the other set includes 20 samples and was used for condition identification test. Figure 8 shows the original vibration signal in each state, and Figure 9 shows the vibration signal after ASTF.

In this study, the SPs that contain the most information and have high sensitivity for each state are selected by PCA. As an example, parts of the selection results are shown in Table 1 and Table 2; $P_1$, $P_2$, $P_6$, $P_8$ and $P_9$ have high sensitivity for distinguishing normal state and abnormal state. Because the weight coefficients for $P_1$, $P_2$, $P_6$, $P_8$ and $P_9$, the first principal component, are larger than those of the other, the contribution rate of the first principal component is larger than 0.86, which contains enough information and discriminatory features to identify the normal state and abnormal state. Similarly, the SPs for other states can also be selected.

In this paper, the DBN for distinguishing conditions of a rolling bearing was built as shown in Figure 10. The proposed DBN consists of three layers. The first layer is normal and abnormal states. The second layer is main failures such as outer-race defect, inner-race defect, and roller element defect, which often occurred in a rolling bearing. The last layer is SPs

**Table 8.** Diagnosis results of NN.

| $P_1$ | $P_2$ | $P_3$ | $P_4$ | $P_5$ | $P_6$ | $P_7$ | $P_8$ | $P_9$ | $P_{10}$ | N | O | I | R | Judge |
|---|---|---|---|---|---|---|---|---|---|---|---|---|---|---|
| 0.02 | 0.06 | 0.13 | 6.13 | 2.24 | 1.0 | 1.98 | 4.72 | 2.01 | 11.9 | 0.868 | 0.001 | 0.135 | 0.012 | N |
| 0.02 | 0.07 | 0.11 | 5.87 | 2.58 | 1.03 | 1.75 | 5.73 | 1.15 | 10.1 | 0.895 | 0.001 | 0.098 | 0.169 | N |
| 0.03 | 0.16 | 0.15 | 103 | 4.85 | 2.01 | 2.33 | 66.3 | 3.05 | 105 | 0.063 | 0.805 | 0.177 | 0.055 | O |
| 0.03 | 0.17 | 0.19 | 106 | 4.62 | 2.15 | 2.68 | 72.5 | 3.66 | 99.7 | 0.087 | 0.796 | 0.206 | 0.036 | O |
| 0.02 | 0.11 | 0.25 | 33.4 | 2.91 | 3.06 | 10.8 | 80.7 | 8.49 | 109 | 0.056 | 0.071 | 0.532 | 0.405 | × |
| 0.04 | 0.09 | 0.32 | 48.9 | 3.65 | 1.95 | 9.67 | 85.6 | 8.98 | 94.9 | 0.095 | 0.041 | 0.501 | 0.386 | × |
| 0.05 | 0.32 | 0.47 | 93.9 | 5.29 | 3.57 | 8.87 | 115 | 10.8 | 124 | 0.011 | 0.095 | 0.406 | 0.513 | × |
| 0.06 | 0.26 | 0.68 | 106 | 5.47 | 3.26 | 9.13 | 123 | 10.2 | 139 | 0.032 | 0.086 | 0.366 | 0.572 | × |
| ⋮ | ⋮ | ⋮ | ⋮ | ⋮ | ⋮ | ⋮ | ⋮ | ⋮ | ⋮ | ⋮ | ⋮ | ⋮ | ⋮ | ⋮ |

shown in Table 2. The prior probabilities and the conditional probabilities were obtained by Equations (35) and (36). All of the SPs were divided into five levels, 1, 2, 3, 4 and 5, which indicate very small, small, middle, big and very big levels, respectively. As an example, parts of the training sample data are shown in Table 3.

To verify the diagnostic capability of the diagnosis methods proposed in this paper, we used the data measured in each state had not been used to train the DBN system. They can correctly and quickly diagnose those faults with the possibility grades of the corresponding states. In the test of normal state, the successful diagnosis ratio is 100%. In the test of each faults, the successful diagnosis ratio of outer-race defect, inner-race defect and roller element defect states are 100%, 94% and 86%, respectively. Some diagnosis results are shown in Table 4, Table 5, Table 6 and Table 7.

## 5.2. Condition Diagnosis by NN

In this study, a back propagation NN shown in Figure 11 is also constructed for condition diagnosis of the roller bearing. The NN consists of three layers, the SPs calculated by vibration signals are entered into input layer, hidden layer includes 80 units, output layer is the possibility grades of each condition of roller bearing. Table 8 shows the parts of diagnosis results of NN. N, O, I and R indicate the normal, outer race defect, inner race defect and roller element defect states, respectively. The symbol × expresses the case that NN is incapable of identifying the fault type. As shown in Table 8, the normal and outer race defect states of roller bearing were correctly identified by NN. However, NN is incapable of identifying inner race defect and roller element defect states of roller bearing. The main reasons are that the vibration signals measured for condition diagnosis contain strong noise, there exist ambiguous relationships between the SPs and the fault types, and NN cannot deal with incomplete and conflicting information.

## 6. CONCLUSIONS

In order to detect the condition of rotating machinery at an early stage, a novel fault diagnosis method based on ASTF and DBN was presented. The main conclusions of this paper are summarized as follows:

The method of ASTF for extracting weak fault features under background noise was presented. The optimal level of significance α was obtained using PSO. To evaluate the performance of ASTF, evaluation factor $I_{pq}$ was also defined. In addition, a simulation experiment was designed to verify the effectiveness and robustness of ASTF.

PCA based on statistical analysis theory was also presented to evaluate the sensitivities of SPs calculated via vibration signals measured in each state for condition identification.

A three-layer DBN was developed to identify condition of rotation machinery based on the BBN theory. It is effective and efficient in condition diagnosis based on uncertain, incomplete and conflicting information.

Study examples of diagnosis for a bearing were shown to demonstrate the effectiveness of the methods proposed in this paper. The verification results show that the bearing faults that often occur in roller bearings, such as the Outer race, the Inner race and the roller element defects, have been effectively identified by the proposed method in this paper. However, these bearing faults are difficult to detect using NN technology, because the vibration signals measured for condition diagnosis contain strong noise, there exist ambiguous relationships between the SPs and the fault types, and NN cannot deal with incomplete and conflicting information.

In summary, this paper verifies the capability of condition diagnosis method based on ASTF and DBN. In addition, soft sensor technique establish inference model of symptom parameters based on state-space model, and solves inference model of symptom parameters by parameter estimation method, such as Kalman filter and Bayesian filter. The condition diagnosis system includes two parts: feature extraction and condition identification. Soft sensor technique can be used of feature extraction and fusion, and condition identification can adopt artificial intelligence techniques, such as DBN, NN, *etc*. In the future, we will consider using soft sensor technique to extract features of signals.

## ACKNOWLEDGMENTS

The authors would like to acknowledge the financial support of the National Natural Science Foundation of China (51575236), Fundamental Research Funds for the Central Universities (Grant No. JUSRP51511). This work was also supported by the open project of Jiangsu key laboratory of advanced food manufacturing equipment & technology (Grant No.FM-2015-01).

## AUTHOR CONTRIBUTIONS

Ke Li, Peng Chen and Huaqing Wang conceived and designed the experiments; Ke Li and Peng Chen performed the experiments; Ke Li, Qiuju Zhang and Kun Wang analyzed the data; Ke Li wrote the paper.

## CONFLICTS OF INTEREST

The authors declare no conflict of interest.

## REFERENCES

1. Zhu, Q.B. Gear fault diagnosis system based on wavelet neural networks. *Dyn. Contin. Discret. Impuls. Syst. Ser. A Math. Anal. Part 2 Suppl.* **2006**, *13*, 671–673.
2. Li, K.; Chen, P.; Wang, S.M. An intelligent diagnosis method for rotating machinery using least squares mapping and a fuzzy neural network.*Sensors* **2012**, *5*, 5919–5939.
3. Mitoma, T.; Wang, H.; Chen, P. Fault diagnosis and condition surveillance for plant rotating machinery using partially-linearized neural network. *Comput. Ind. Eng.* **2008**, *55*, 783–794.
4. Pacas, M.; Villwock, S.; Dietrich, R. Bearing damage detection in permanent magnet synchronous machines. In Proceedings of 2009 Energy Conversion Congress and Exposition, San Jose, CA, USA, 24–29 September 2009; pp. 1098–1103.
5. Stack, J.R.; Habetler, T.G.; Harley, R.G. Fault-signature modelingand detection of inner-race bearing faults. In Proceedings of the IEEE International Conference on Electric Machines and Drives, San Antonio, TX, USA, 15 May 2005.
6. Gao, Z.W.; Cecati, C.; Ding, S.X. A survey of fault diagnosis and fault-tolerant techniques part I: Fault diagnosis with model-based and signal-based approaches. *IEEE Trans. Ind. Electron.* **2015**, *62*, 3757–3767.
7. Gao, Z.W.; Cecati, C.; Ding, S.X. A real-time fault diagnosis and fault-tolerant control. *IEEE Trans. Ind. Electron.* **2015**, *62*, 3752–3756.

8. Feng, Z.; Zuo, M. Fault diagnosis of planetary gearboxes via torsional vibration signal analysis. *Mech. Syst. Process.* **2013**, *36*, 401–421.
9. Zhou, Y.; Han, J.; Li, Z.N.; Qu, H.T. Fractional Fourier transform and application to rolling bearing fault diagnosis. *Coal Mine Mach.* **2007**.
10. Rajagopalan, S.; Restrepo, J.A.; Aller, J.M.; Habetler, T.G.; Harley, R.G. Nonstationary motor fault detection using recent quadratic time-frequency representations. *IEEE Trans. Ind. Appl.* **2008**, *5–6*, 735–744.
11. Randall, R.B.; Antoni, J. Rolling element bearing diagnostics—A tutorial. *Mech. Syst. Signal Process.* **2011**, *25*, 485–520.
12. Long, J.; Wu, J.Q. Application of Short Time Fourier Transform and Hilbert-Huang Transform in Fault Diagnosis of Rolling Bearings of Windmill. *Noise Vib. Control* **2013**, *33*, 219–222.
13. Luo, X.P.; Du, P.Y. A fault diagnosis method of motor based on wavelet transform. In Proceedings of the Second IITA International Conference on Geoscience and Remote Sensing, Qingdao, China, 28–31 August 2010.
14. Ordaz-Moreno, A.; Romero-Troncoso, R.J.; Vite-Frias, J.A.; Rivera-Gillen, J.R.; Garcia-Perez, A. Automatic online diagnosis algorithm for broken-bar detection on induction motors based on discrete wavelet transform for FPGA implementation. *IEEE Trans. Ind. Electron.* **2008**, *55*, 2193–2202.
15. Sheen, Y.T. An analysis method for the vibration signal with amplitude modulation in a bearing system. *J. Sound Vib.* **2007**, *303*, 538–552.
16. Bin, G.F.; Gao, J.J.; Li, X.J.; Dhillon, B.S. Early fault diagnosis of rotating machinery based on wavelet packets—Empirical mode decomposition feature extraction and neural network. *Mech. Syst. Signal Process.* **2012**, *27*, 696–711.
17. Wu, F.J.; Qu, L.S. Diagnosis of subharmonic faults of large rotating machinery based on EMD. *Mech. Syst. Signal Process.* **2009**, *23*, 467–475.
18. Huang, N.E.; Shen, Z.; Long, S.R.; Wu, M.L.C.; Shih, H.H.; Zheng, Q.N. The empirical mode decomposition and the Hilbert spectrum for nonlinear and non-stationary time series analysis. *Proc. Royal Soc. A Math. Phys. Eng. Sci.* **1998**, *454*, 903–995.

19. Boudraa, A.O.; Cexus, J.C. EMD-based signal filtering. *IEEE Trans. Instrum. Meas.* **2007**, *56*, 2196–2202.
20. Tandon, N.; Choudhury, A. A review of vibration and acoustic measurement methods for the detection of defects in rolling element bearings.*Tribol. Int.* **1999**, *32*, 469–480.
21. Volponi, A.J.; DePold, H.; Ganguli, R.; Chen, D. The use of Kalman filter and neural network methodologies in gas turbine performance diagnostics: A comparative study. *J. Eng. Gas Turbines Power* **2003**, *125*, 917–924.
22. Gardner, W.A. Cyclic Wiener Filtering: Theory and Method. *IEEE Trans. Commun.* **1993**, *41*, 151–163.
23. Sorsa, T.; Koivo, H.N.; Koivisto, H. Neural networks in process fault diagnosis. *IEEE Trans. Syst. Man Cybern.* **1991**, *21*, 815–825.
24. Li, K.; Wang, H.Q.; Chen, P. Intelligent diagnosis method for rotating machinery using wavelet transform and ant colony optimization. *IEEE Sens. J.* **2012**, *12*, 2474–2484.
25. Pearl, J. Bayesian networks: A model of self-activated memory for evidential reasoning. In Proceedings of the 7th Conference of the Cognitive Science Society, University of California, Irvine, CA, USA, 15–17 August 1985.
26. Pearl, J. Fusion, propagation, and structuring in belief networks. *Artif. Intell.* **1986**, *29*, 241–288.
27. Kang, C.W.; Golay, M.W. A Bayesian belief network-based advisory system for operational availability focused diagnosis of complex nuclear power systems. *Expert Syst. Appl.* **1999**, *17*, 21–32.
28. Sahin, F.; Yavuz, M.; Arnavut, Z.; Uluyol, O. Fault diagnosis for airplane engines using Bayesian networks and distributed particle swarm optimization. *Parallel Comput.* **2007**, *33*, 124–143.
29. Yang, L.; Lee, J. Bayesian belief network-based approach for diagnostics and prognostics of semiconductor manufacturing systems. *Robot. Comput. Integr. Manuf.* **2012**, *28*, 66–74.
30. Kennedy, J.; Eberhart, R. Particle swarm optimization. In Proceedings of IEEE International Conference on Neural Networks, the University of Western Australia, Perth, Australia, 27 November–1 December 1995; pp. 1942–1948.

31. Shao, P.; Wu, Z.J. Rosenbrock function optimization based on improved particle swarm optimization algorithm. *Comput. Sci.* **2013**, *40*, 194–197.
32. Akjiratikarl, C.; Yenradee, P.; Drake, P.R. PSO-based algorithm for home care worker scheduling in the UK. *Comput. Ind. Eng.* **2007**, *53*, 559–583.
33. Fukunaga, K. *Introduction to Statistical Pattern Recognition*; Academic Press: San Diego, CA, USA, 1972.
34. Lu, N.; Wang, F.; Gao, F. Combination method of principal component and wavelet analysis for multivariate process monitoring and fault diagnosis. *Ind. Eng. Chem. Res.* **2003**, *42*, 4198–4207.

# CHAPTER 3

## BAYESIAN MARKOV REGIME-SWITCHING MODELS FOR COINTEGRATION

Kai Cui[1], Wenshan Cui[2]

[1]Department of Statistical Science, Duke University, Durham, USA

[2]School of Science and Information, Qingdao Agricultural University, Qingdao, China

### ABSTRACT

This paper introduces a Bayesian Markov regime-switching model that allows the cointegration relationship between two time series to be switched on and off over time. Unlike classical approaches for testing and modeling cointegration, the Bayesian Markov switching method allows for estimation of the regime-specific model parameters via Markov Chain Monte Carlo and generates more reliable estimation. Inference of regime switching also provides important information for further analysis and decision making.

### KEYWORDS

Cointegration; Regime-Switching; Bayesian; MCMC

### 1. INTRODUCTION

Since the development of the concept of cointegration [1], there has been a rich literature on testing cointegration and applying cointegration approaches to real data analysis. One of the most illustrative examples in practice is the pair trading strategy [2]. The basic idea is that: find two securities whose prices have been historically moving together. So when the spread between them widens, we short the winner and buy the loser. And if we believe that the history would repeat itself, prices will converge again and the arbitrager will profit. This moving-together relationship between two nonstationary time series is called cointegration. Mathematically, if two nonstationary time series $U_t$ and $V_t$ are cointegrated, then there exists a number $\delta$ called the cointegration ratio, such that $Y_t = U_t - \delta V_t$ is stationary.

Although there have been many statistical studies to find cointegrated time series, there are still many unsolved problems. First of all, it is often hard simply to find cointegration given a specific period of time. There are several statistical explanations for failing to reject the null of no cointegration including the span of the data set, structural breaks [3] and the choice of test model [4]. Secondly, there are few statistical decision-making rules after identifying candidate pairs. Taking pair trading as an exmple, typically, people simply use the decision rule that they open a long-short position when the pair prices have diverged by a certain amount (e.g. two standard deviations from the historical mean) and close the position when the prices have reverted [5].

This paper proposes the Bayesian Markov regime switching model that allows the cointegration relationship between two time series to be switched on or off over time via a discrete-time Markov process. This is an improvement to the traditional cointegration tests considering that the model flexibly allows local non-cointegration rather than assuming global cointegration over the whole period of time. By using a fully Bayesian models, uncertainty about cointegration ratio is also incorporated into the model and inferred simultaneously with all other unknown quantities. Furthermore, inference of the hidden regime-switching is also critical to decision making and further generic analysis.

## 2. MARKOV REGIME-SWITCHING MODELS FOR COINTEGRATION

Suppose we have two nonstationary time series $U_t$ and $V_t$ with integration order 1, and $Y_t = U_t - \delta V_t$ ($\delta$ is known, typically people propose a $\delta$ and then test the stationary property of $Y_t$). If $Y_t$ is stationary, then we say time series $U_t$ and $V_t$ are cointegrated. To test for stationarity, the Engle-Grange method [6] tests the $\gamma = 0$ null hypothesis using the ADF unit root test [5] based on the Error Correction Model (EVM) with lag order K (as compared to $\gamma < 0$ in which case it is stationary):

$$\Delta Y_t = \mu + \gamma Y_{t-1} + \sum_{k=1}^{K} \beta_i \Delta Y_{t-i} + \varepsilon_t \qquad (1)$$

where $\mu$ is a constant, $\beta_i$ s are autoregression coefficients and $\varepsilon_t \sim N(0, \Sigma_\varepsilon)$.

In comparison, the Markov regime-switching model we proposed allows $Y_t$ to switch between cointegrated or non-cointegrated regimes in a Markovian manner, by introducing the regime indicator variable $X_t$, regime specific parameters and the Markov transition matrix $P_X$. For the simplicity of exposition, we assume that $X_t \in \{0,1\}$, with $X_t = 0$ denoting that that $Y_t$ is stationary (i.e. $U_t$ and $V_t$ are cointegrated) at time t and $X_t = 1$ meaning non-cointegration. Then the model can be written as:

$$\Delta Y_t = \mu^{(X_t)} + \gamma I_{\{X_t = 0\}} Y_{t-1} + \sum_{k=1}^{K} \beta_k^{(X_t)} \Delta Y_{t-i}$$
$$+ \varepsilon_t, \varepsilon_t \sim N\left(0, \Sigma_\varepsilon^{(X_t)}\right) \qquad (2)$$

$$P_X = \begin{bmatrix} p_{00} & p_{01} \\ p_{10} & p_{11} \end{bmatrix}$$

where $Y_t = U_t - \delta V_t$ and thus $\Delta Y_t = \Delta U_t - \delta \Delta V_t$. $P_X$ is the Markov transition matrix of $X_t$, with $p_{ij} = \Pr(X_{t+1} = j | X_t = i)$ and initial value $X_0$.

Clearly, when $X_t = 0$ the model reduces to model (1) with negative $\gamma$, while $X_t = 1$ specifies unit root process for $Y_t$ and thus no cointegration exists for time series $U_t$ and $V_t$. By obtaining inference of the underlining regimes $X_t$, regime-specific parameters and segmentation of regime-specific data, the model provides much information for further generic analysis and decision making.

## 3. BAYESIAN COMPUTATION

We propose to use Bayesian analysis for the inference of parameters and latent regimes $X_t$, where posterior samples of all unknown quantities are drawn using Markov Chain Monte Carlo (MCMC).

Under this model (2), the likelihood function is:

$$L(X,Y|\Theta) \propto P(X_0)P(\Delta Y_0|X_0)$$

$$\cdot \left[ \prod_{t=1} P(\Delta Y_t|X_t)P(X_t|X_{t-1}) \right]$$

$$\propto P(X_0|)P(\Delta Y_0|X_0)$$

$$\left[ \prod_{t:X_t=0} P(\Delta Y_t|X_t)P(X_t|X_{t-1}) \right]$$

$$\cdot \left[ \prod_{t:X_t=1} P(\Delta Y_t|X_t)P(X_t|X_{t-1}) \right]$$

Conjugate prior distributions are placed on model parameters [7]. Specifically, conjugate Dirichlet priors are assigned to each row of the transition matrix $P_X$ and $q = (q_0, q_1)$, where $q_i = \Pr(X_0 = i)$. Conjugate Normal-Gamma priors are assigned for all the regression coefficients $\beta_k^{(X_t)}$ and the corresponding precisions $\phi_\varepsilon^{(X_t)}$

$$q \sim Dir(\alpha_1, \alpha_2); p_{ij} \sim Dir(\alpha_1^{(i)}, \alpha_2^{(i)}); i, j = 0,1$$

$$\beta_k^{(X_t)} \sim N\left(0, \frac{\lambda}{\phi^{(X_t)}}\right); \phi^{(X_t)} \sim \Gamma\left(\frac{v_0}{2}, \frac{SS_0}{2}\right)$$

To obtained the posterior marginal distributions of the unknown parameters and the hidden regimes $X_t$, Gibbs sampler is constructed to iterate the following steps:

1) Sample $q$ and $P_X$ from full conditional distributions:

$$q|\cdots \sim Dir\left(\alpha_1 + I_{\{X_0=0\}}, \alpha_2 + I_{\{X_0=1\}}\right); (p_{00}, p_{01})|\cdots$$

$$\sim Dir\left(\alpha_1^{(0)} + \sum_{t=1} I_{\{X_{i-1}=0, X_i=0\}}, \alpha_2^{(0)} + \sum_{t=1} I_{\{X_{i-1}=0, X_i=1\}}\right)$$

$$\cdot (p_{10}, p_{11})|\cdots$$

$$\sim Dir\left(\alpha_1^{(1)} + \sum_{t=1} I_{\{X_{i-1}=1, X_i=0\}}, \alpha_2^{(1)} + \sum_{t=1} I_{\{X_{i-1}=1, X_i=1\}}\right)$$

2) Sample the regression coefficients $\beta$ and variance from Normal-Inverse Gamma full conditional distributions given the conjugacy of the priors.

3) Sample the whole path of $X_t$.

Since $X_t$ are highly correlated, Gibbs sampler constructed via regular full conditional distribution would be extremely inefficient [7]. To overcome this, Forward Filtering and Backward Sampling algorithm is applied to draw block samples of $X_t$. To achieve this, define $\alpha_n(i) = P(\Delta Y_0, \cdots, \Delta Y_n, X_n = i | \Theta)$, then by recursion:

$$\alpha_0(i) = q_i P(\Delta Y_0 | X_0 = i);$$

and

$$\alpha_n(i) = \left[\sum_j \alpha_{n-1}(j) p_{ji}\right] P(\Delta Y_n | X_n = i)$$

With this, the results follow that:

$$P(X_t = i | \cdots) \propto \alpha_t(i)$$
$$P(X_{t-1} = j | X_t = i, \cdots) \propto \alpha_{t-1}(j) p_{ji}$$

By using this algorithm, a sample of $X_T$ is first drawn from a bernoulli (multinomial if $X_t$ takes more than two values) distribution, and $\forall 0 \leq t \leq T-1$, samples of $X_t$ are drawn sequentially and backward from the conditional bernoulli distribution, with

$$P(X_t = j | X_{t+1} = i) = \frac{\alpha_t(j) p_{ji}}{\sum_j \alpha_t(j) p_{ji}}$$ until the whole $X_t$ time series are sampled.

## 4. SIMULATED TIME SERIES ANALYSIS

### 4.1. Model Assessment

To testify the performance of the proposed framework, we simulated a Markov regime-switching times series of length $T = 500$, which switches between one stationary AR(2) process (State 0) and one non-stationary AR(2) process (State 1). The two AR(2) models and the corresponding Error Correction Models (ECM) are shown as follows: (the (non-)stationary property can be easily tested by the Unit Root Test)

$$\begin{aligned}
&\text{State 0}: y_t = 3 + 0.6 y_{t-1} - 0.28 y_{t-2} + \varepsilon_t, \\
&ECM: \Delta y_t = 3 - 0.68 y_{t-1} + 0.28 \Delta y_{t-1} + \varepsilon_t; \\
&\text{State1}: y_t = 4 + 0.7 y_{t-1} + 0.3 y_{t-2} + \varepsilon_t^*, \\
&ECM: \Delta y_t = 4 - 0.3 \Delta y_{t-1} + \varepsilon_t^*.
\end{aligned} \quad (3)$$

where $\varepsilon_t \sim N(0,1)$ and $\varepsilon_t' \sim N(0,1)$. The transition matrix is specified as

$$P_X = \begin{bmatrix} p_{00} & p_{01} \\ p_{10} & p_{11} \end{bmatrix} = \begin{bmatrix} 0.8 & 0.2 \\ 0.8 & 0.2 \end{bmatrix}$$

A simulated data was shown in **Figure 1**.
The proposed model was applied to the time series to find regime switching, with the priors specified as follows:

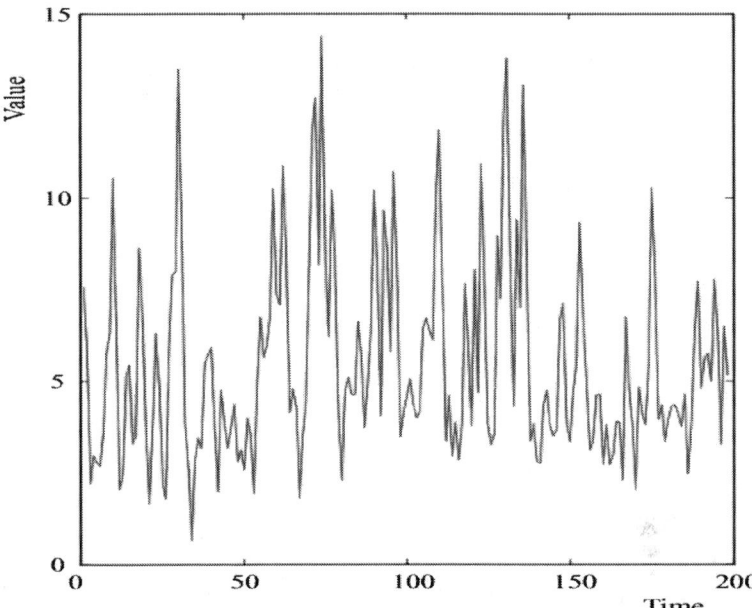

**Figure 1**. Illutration of a time series stimulated by the markov switching model.

$$q \sim Dir(1,1), (p_{00}, p_{01}) \sim Dir(1,1), (p_{10}, p_{11}) \sim Dir(1,1),$$

$$\beta_k^{(X_t)} \sim N\left(0, \frac{1}{\phi^{(X_t)}}\right); \phi^{(X_t)} \sim \Gamma\left(\frac{1}{2}, \frac{1}{2}\right)$$

To infer the value of $X_t$ based on posterior samples, we use posterior probability $> 0.5$ as the cut-off point. Shown in **Figure 2**, the inferred regimes are compared with the true values, which shows that our model gives good recovery of the latent regimes (with the first 200 time points shown). Other model parameters are also correctly inferred as shown in **Table 1**, where posterior distributions cover the true values well.

## 4.2. Posterior Decision Making

The importance of inference of regimes when analyzing (non)stationary time series lies in the fact that commonly-used stationarity and cointegra-

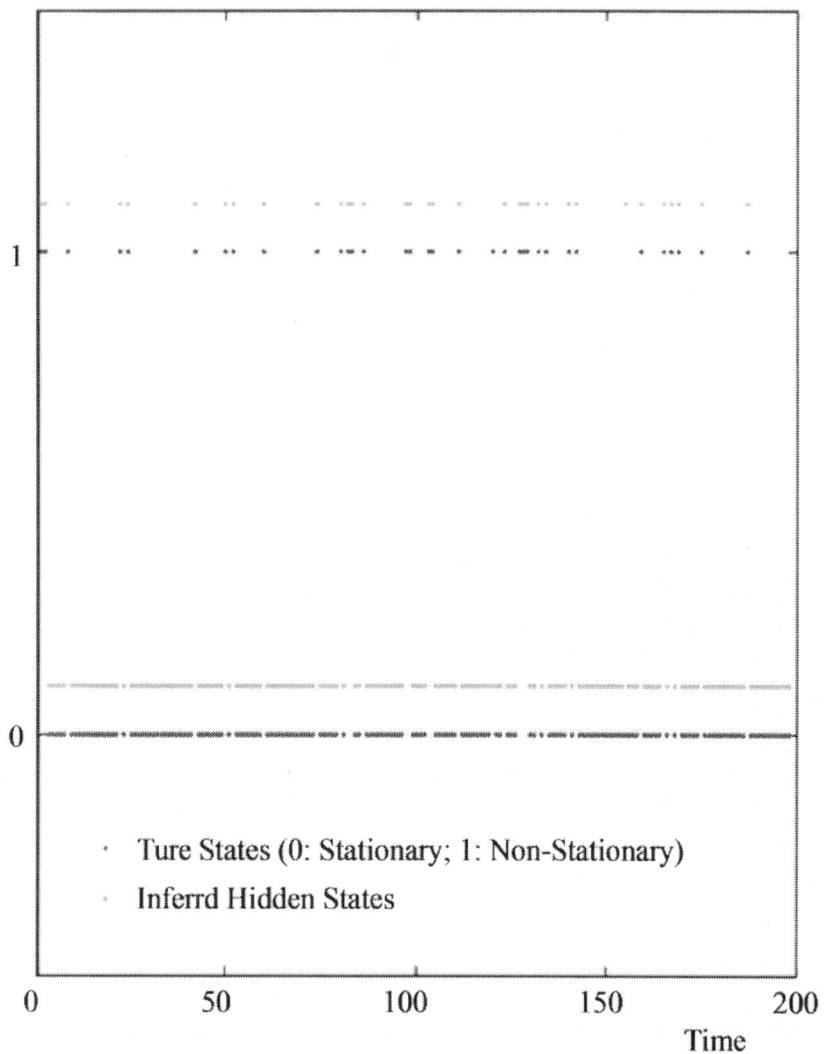

**Figure 2.** Inferred regimes $X_t$ (in green) compared to the true values (in blue) show good inference.

tion tests (e.g. ADF unit root test and Engle-Granger cointegration tes [1]) may well give misleading results when regime switching exists in the process. For illustration, a quick ADF test of the previously simulated data concludes that the null hypothesis with unit root is rejected at 99.9% confidence level, indicating the times series is stationary. If this time series were generated by the linear combination of two nonstationary time series, then the ADF test tells that these two are co-integrated, which is clearly wrong.

**Table 1.** Posterior estimates of model parameters compared to the true values. The parameters are defined as in model 2 and specified in (3).

| Parameter | Mean | STD | Truth |
|---|---|---|---|
| $\gamma$ | −0.676 | 0.011 | −0.68 |
| $\beta_1^{(0)}$ | 0.287 | 0.013 | −0.28 |
| $\phi^{(0)}$ | 0.952 | 0.048 | 1 |
| $\beta_1^{(1)}$ | −0.323 | 0.041 | −0.3 |
| $\phi^{(1)}$ | 0.911 | 0.093 | 1 |

In the following part, we will use the context of pair trading to illustrate how the Markov regime-switching model can potentially help improve decision making in practice. Basically people do pair trading based on the traditional rule that you open a long-short position when the pair prices have diverged by more than two historical standard deviations. And you unwind the position when it returns to historical mean.

First of all, the model clearly allows more reasonable estimation of the historical mean and standard deviation, based soly on data in the stationary (cointegrated) regimes, rather than including data in the nonstationary (non-cointegrated) regimes. This difference can be observed in **Figure 3**, where the historical mean using data in the stationary regime is different from that using all data, and the standard deviation is also smaller.

Secondly, the identification of stationary (cointegrated) and nonstationary (non-cointegrated) regimes also help establish more rational decision making rules, which should be: we open a position when it is both in the stationary state and has diverged from the historical stationary mean. It is apparently risky either to open a position when currently we are in a non-stationary state or the historical mean calculation involves non-stationary data.

Since people care much about the time points where values are at least 2 standard deviations away from the historical mean, the figure shows that the we pick different time points using our model and decision making rules from those obtained using all historical data and traditional rules, which we believe are more reasonable choices. For example, many spikes in **Figure 3** are actually not good time points to open the position based on

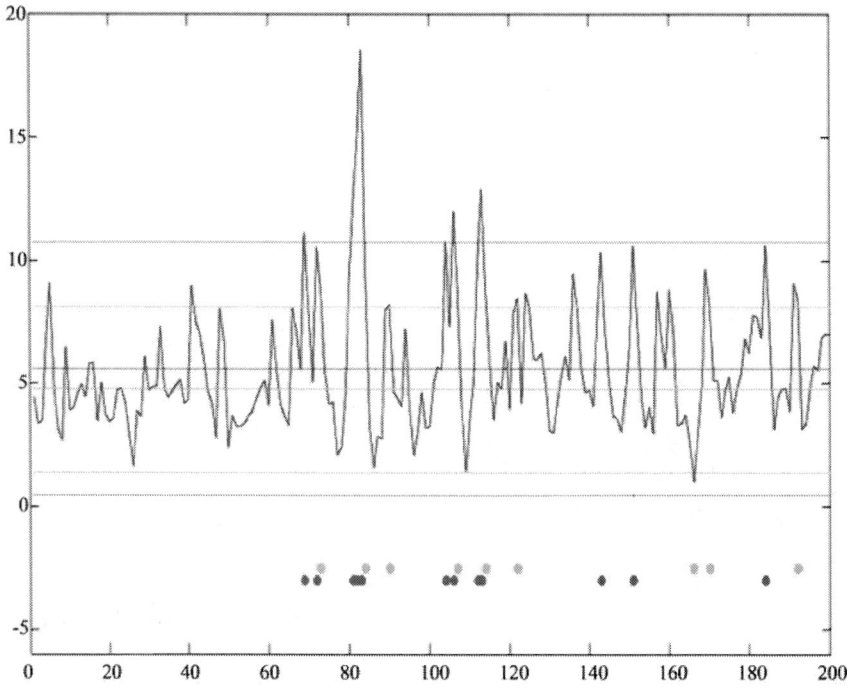

**Figure 3**. Results comparison between our Bayesian Markov regime-switching model and traditional cointegration test and analysis using all historical data. Red lines indicate the mean and mean ±2SD using all historical data, which is a traditional way after you have done the ADF test to show the stationary property; Green lines indicate those using only historical data in stationary regimes. Red and green dots mark the time points where values at those points are at least 2SD away from the historical mean based on traditional and our Markov regime switching model respectively.

our Markov regime-switching model simply because those spikes are in the non-stationary (non-cointegrated) regime. However in comparison, the traditional approach considers them open positions whenever the values are 2 standard deviations away from the mean, which is a very risky decision not considering the regimes.

## 5. COINTEGRATED PRICE SERIES ANALYSIS

An possible example of a pair of cointegrated time series is the gold ETF, GLD versus the gold miners ETF, GDX. GLD reflects the spot price of gold,

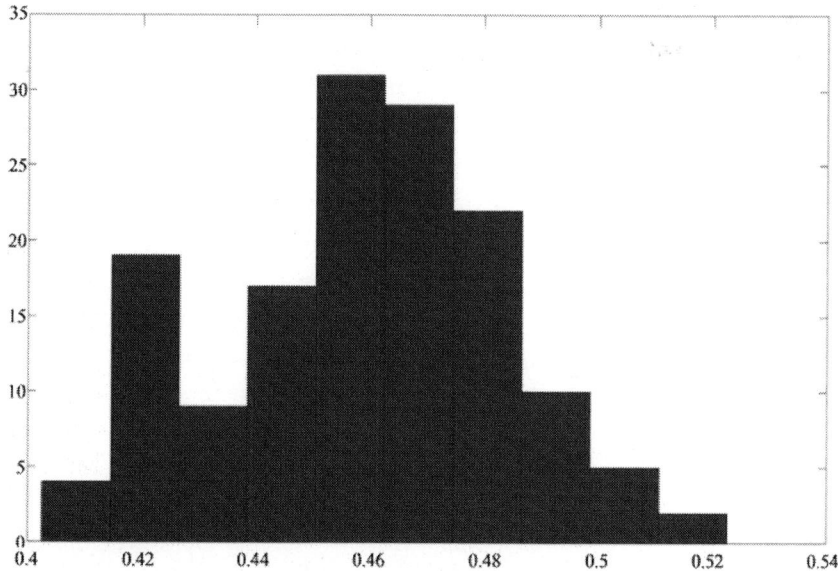

**Figure 4**. Distribution of the probability that $X_t$ is in the cointegration regime (t = 1,···,T).

and GDX is a basket of gold-mining stocks. It makes intuitive sense that their prices may move in tandem. Previous study via the two step Engle-Granger method [1] identified that a portfolio with long 1 share of GLD and short 1.6766 share of GDX is likely a stationary time series, with lag 1 but the conclusion is later questioned by other studies [8]. To test the possible co-integration, the two-state Markov regime switching model is applied to the 05/23/06-11/30/07 GLD and GDX time series. A histogram shown the distribution of the probability of the time points being in the cointegration state is shown in **Figure 4**. According to the previous 0.5 cut-off point, the Markov regime switching model indicates that at most of the time, the two time series are not cointegrated with the 1.6766 cointegration ratio. This may serve as another counterexample (together with the simulation result) that the widely-used ADF test might provide misleading results when used to test co-integration regardless of possible regime switching.

## 6. CONCLUSIONS AND FUTURE WORK

In this study, we proposed to use the Bayesian Markov regime-switching model as a flexible model for cointegration and stationarity analysis, where the latent regimeswitching process is modeled via a Markov process. A strong message of this study is that, while identifying cointegration (or stationarity) is often hard globally, allowing local non-cointegration (or non-stationarity) and inferring the regime switching can provide much information for further analysis and decision making.

Several extensions of the study are still worth exploring, including relaxing the hidden Markov transition models and incorporating uncertainty about number of regimes in the model. Hidden semi-Markov models are natural extensions of hidden Markov models. While the runlength distribution of the hidden Markov models implicity follows a geometric distribution, hidden semi-Markov models allow for more general runlength distributions, and thus are more flexible to describe the time spend in a given regime. As for the cases with the number of regimes unknown, Bayesian inference through reversible jump MCMC methods [9] could be a viable alternative that both explores models with different number of regimes and estimation of regime-specific parameters.

## REFERENCES

1. R. F. Engle and C. W. J. Granger, "Co-integration and Error Correction: Representation, Estimation, and Testing," Econometrica, Vol. 55, No. 2, 1987, pp. 251-276.
2. H. Puspaningrum, "Pairs Trading Using Cointegration Approach," Ph.D. Thesis, University of Wollongong, Wollongong, 2012.
3. J. Campos and N. R. Ericsson and D. F. Hendry, "Cointegration Tests in the Presence of Structural Breaks," International Finance Discussion Papers 440, Board of Governors of the Federal Reserve System (US), 1996.
4. E. G. Gatev and W. Goetzmann and K. G. Rouwenhorst, "Pairs Trading: Performance of a Relative Value Arbitrage Rule," Boston College, Boston, 2006.
5. A. W. Gregory and B. E. Hansen, "Residual-Based Tests for Cointegration in Models with Regime Shifts," Journal of Econometrics, Vol. 70, No. 1, 1996, pp. 99-126.
6. D. A. Dickey and W. A. Fuller, "Distribution of the Estimators for Autoregressive Time Series With a Unit Root," Journal of the American Statistical Association, Vol. 74, No. 366, 1979, pp. 427-431.

7. G. E. B. Archer and D. M. Titterington, "Parameter Estimation for Hidden Markov Chains," Journal of Statistical Planning and Inference, Vol. 108, No. 1, 2002, pp. 365- 390.
8. E. P. Chan, "Quantitative Trading," John Wiley and Sons, Hoboken, 2008.
9. C. P. Robert, T. Ryden and D. M. Titterington, "Bayesian Inference in Hidden Markov Models through the Reversible Jump Markov Chain Monte Carlo Method," Journal of The Royal Statistical Society Series B, Vol. 62, No. 1, 2000, pp. 57-75.

# CHAPTER 4

## CLASSIFICATION OF WEB SERVICES USING BAYESIAN NETWORK

Ramakanta Mohanty[1], V. Ravi[2], M. R. Patra[3]

[1]Computer Science Department, MLR Institute of Technology, Hyderabad, India;

[2]Institute for Development and Research in Banking Technology, Hyderabad, India;

[3]Computer Science Department, Berhampur University, Berhampur, India.

### ABSTRACT

In this paper, we employed Naïve Bayes, Markov blanket and Tabu search to rank web services. The Bayesian Network is demonstrated on a dataset taken from literature. The dataset consists of 364 web services whose quality is described by 9 attributes. Here, the attributes are treated as criteria, to classify web services. From the experiments, we conclude that Naïve based Bayesian network performs better than other two techniques comparable to the classification done in literature.

### KEYWORDS

Web Services; Quality of Services (QoS); Bayesian Network; Naïve Based Bayesian; Markov Blanket and Tabu Search

## 1. INTRODUCTION

Services are tendered and availed in almost all the business and industries. The growth and proliferation of IT across industries and business appear to have fuelled the requirement as well as the delivery of services profoundly. Delivering services has been an attractive business proposition for many industries lately. The latest development in the systems is a new paradigm, called web services [1]. Web Services heralded another significant mile stone in the history of IT. Earlier, Internet catered mostly to the business to Customer (B2C) category of the users on the web. As against this, Web Services enable B2B interaction as well Web. They are independent of platform and natural languages, which is suitable for accessing from heterogeneous environments. With the rapid introduction of web services technologies, researchers focused more on the functional and interfacing aspects of web services, which include HTTP and XMLbased messaging. They are used to communicate by employing pervasive standards based web technologies. Web services are based on XML and three other core technologies: WSDL, SOAP, and UDDI. WSDL is a document which describes the services' location on the web and the functionality the service provides. Information related to the web service is to be entered in a UDDI registry, which permits web service consumers to find out and locate the services they required. With the help of the information available in the UDDI registry based on the web services, client developer uses instructions in the WSDL to construct SOAP messages for exchanging data with the service over HTTP attributes [2].

In this study, we address this problem of efficiently identifying a set quality attributes by employing Bayesian Networks vz. Naive Bayes, Markov blanket and Tabu search. Naive Bayes is special form of Bayesian network that is widely used for classification [3] and clustering [4], but its potential for general probabilistic modeling (i.e., to answer joint, conditional and marginal queries over arbitrary distributions) remains largely unexploited. Naive Bayes represents a distribution as a mixture of components, where within each component all variables are assumed independent of each other. The Markov blanket of a variable Y, (MB(Y)), by definition, is the set of variables such that Y is conditionally independent of all the other variables given MB(Y). A Markov Blanket Directed Acyclic Graph (MB DAG) is a Directed Acyclic Graph over that subset of variables. When the parameters of the MB DAG are estimated, the result is a Bayesian Network, defined in the next section. Recent research by the machine learning community [5-7] has sought to identify the Markov blanket of a target variable by filtering variables using statistical decisions for conditional independence and using the MB predictors as the input features of a classifier. However, learning MB DAG classifiers from data is an open problem [8]. There are several challenges: the problem of learning the graphical structure with the highest score (for a variety of scores) is NP hard [8] for methods that use conditional independencies to guide graph search, identifying conditional independencies in the presence of limited data is quite unreliable and the presence of multiple local optima in the Tabu search Enhanced Markov blanket Classifier space of possible structures makes the search process difficult.

Classification using the Markov blanket of a target variable in a Bayesian Network has important properties. It specifies a statistically efficient prediction of the probability distribution of a variable from the smallest subset of variables that contains all of the information about the target variable, it provides accuracy while avoiding over fitting due to redundant variables and it provides a classifier of the target variable from a reduced set of predictors. The TS/MB procedure proposed in this paper allows us to move through the search space of Markov blanket structures quickly and escape from local optima, thus learning a more robust structure.

The rest of paper is organized as follows: Section 2 presents the quality issues in web services and QWS dataset. Section 3 describes the overview of Bayesian Network Section 4 presents the results and discussions, and Section 5 concludes the paper.

## 2. QUALITY ISSUES IN WEB SERVICES

QoS plays an important role in finding out the performance of web services. Earlier, QoS has been used in networking and multimedia applications. Recently, there is a trend in adopting this concept to web services [9]. The basic aim is to identify the QoS attributes [10-12] for improving the quality of web services through replication services [10], load distribution [13], and service redirection [14]. To measure the QoS of a web service, attributes like Response Time, Throughput, Availability, Reliability, Cost, and Response Time are considered.

## 2.1. Qos Attributes

According to Kalepu et al. [15], quality of service (QoS) is a combination of several qualities or properties of a service, such as: 1) Availability; 2) Reliability; 3) Price; 4) Throughput; 5) Response Time; 6) Latency; 7) Performance; 8) Security; 9) Regulatory; 10) Accessibility; 11) Robustness/Flexibility; 12) Accuracy; 13) Servability; 14) Integrity and 15) Reputation. QoS parameters determine the performances of the web services and find out which web services are best and meet user's requirements.

Users of web services are not human beings but programs that send requests for services to web service providers. QoS issues in web services have to be evaluated from the perspective of the providers of web services (such as the airline-booking web service) and from the perspective of the users of these services (in this case, the travel agent site) [16]. There are other models available related to the quality of web services issues. A QoS model [16] represented in **Table 1** shows that the main classification of QoS attributes is based on internal attributes, which are independent of the service environment, and external attributes that are dependent on the service environment. The attributes of the model in **Table 1** are almost similar to the attributes of QWS Dataset used in this paper.

**Table 1.** QoS model of web services [4].

| QOS Factor | Internal attributes (Metrics) | External Attributes (Metrics) |
|---|---|---|
| Reliability | Correctness (accuracy and precision) | Availability and consistency |
| Performance | Efficiency (Time and Space Complexity) | Load management (Throughput, waiting and response time security |
| Integrity | - | Security |
| Usability | Input and output attributes | - |

## 2.2. Description of Qws Dataset

QWS dataset [17] consists of different web service implementations and their attributes as presented in **Table 2**. The classification is measured based on the overall quality rating provided by all the attributes. The functionality of the web services can be helpful to differentiate between various services. The attributes G1 to G10 are used as explanatory variables and the attribute

**Table 2**. Attributes of QWS dataset.

| ID | Attribute Name | Description | Units |
|---|---|---|---|
| $G_1$ | Response Time | Time taken to send a request and receive a response | ms |
| $G_2$ | Availability | Number of successful invocations/total invocations | % |
| $G_3$ | Throughput | Total number of invocations for a given period of time | Invokes/second |
| $G_4$ | Successability | Number of Response/Number of request messages | % |
| $G_5$ | Reliability | Ratio of the number of error messages to total messages | % |
| $G_6$ | Compliance | The extent to which a WSDL document follows WSDL Documentation | % |
| $G_7$ | Best Practices | The extent to which a web service follows | % |
| $G_8$ | Latency | Time taken for the server to process a given request | ms |
| $G_9$ | Documentation | Measure a documentation (i.e. description tags) in WSDL | % |
| $G_{10}$ | WSRF | Web service relevance function: a rank for web service Quality | % |

G11 is used as the target variable. However, attributes G12 and G13 are ignored as they do not contribute to the analysis.

The web services [1,2] in the QWS dataset are classified into four categories, such as: 1) Platinum (high quality); 2) gold; 3) silver and 4) bronze (low quality). The classification is measured based on the overall quality rating provided by WSRF. It is grouped into a particular web service based on classification. The functionality of the web services can be helpful to differentiate between various services [14].

## 3. OVERVIEW OF BAYESIAN NETWORKS

A Bayesian network is a directed acyclic graph model that represents conditional independencies between a set of variables [18,19]. It has two constituents: One is a network graphical structure which is a directed acyclic graph with the nodes of variables and arcs of relations. The other is the conditional probability table associated with each node in the model graph. Machine learning techniques are able to estimate the structure and the conditional probability table from the training data. Based on the Bayesian probability inference, the conditional probability can be estimated from the statistical data and propagated along the links of the network structure to the target label. By setting a threshold of confidence, the final probability value can be used as the indication for the classification decision. The Bayesian formula can be mathematically expressed as below:

$$P(H_J|E) = \frac{P(E|H_J) \times P(H_J)}{\sum_{i=1}^{n} P(E|H_i) \times P(H_i)} \quad (1)$$

$$j = 1, 2, \cdots, n$$

According to the basic statistical theory, e.g., the Chain Rule and independency relation derived from the network structure, the joint probability of $E$ can be calculated by the production of local distributions with its parent nodesi.e.

$$P(E)=\prod_{i=1}^{n}P\big(E_i\big|\text{Parent of }(E_i)\big) \qquad (2)$$

In the above formulas, $E$ denotes a set of variable values, i.e. $E = \{E_1, E_2, \cdots, E_n\}$. H is termed as hypothesis. H is called the prior probability and $P(H|E)$ is called posteriori probability of H given $E$. If $E_i$ has no parent nodes, Parent Of $(E_i)$ is equal to $P(E_i)$.

## 3.1. Naïve Bayes

Naive Bayes models are so named for their "naive" assum-ption that all variables Xi are mutually independent given a "special" variable C. The joint distribution is then given compactly by $P(C, X_1, \cdots, X_n) = P(C)\prod_{i=1}^{n} P(X_i|C)$. The univariate conditional distributions $P(X_i|C)$ can take any form (e.g., multinomial for discrete variables, Gaussian for continuous ones). When the variable C is observed in the training data, naive Bayes can be used for classification, by assigning test example $(X_1, \cdots, X_n)$ to the class C with highest $P(C|X_1, \cdots, X_n)$ [2]. When C is unobserved, data points $(X_1, \cdots, X_n)$ can be clustered by applying the EM algorithm with C as the missing information, each value of C corresponds to a different cluster, and $P(C|X_1, \cdots, X_n)$ is the point's probability of membership in cluster C [14]. Naive Bayes models can be viewed as Bayesian networks in which each Xi has C as the sole parent and C has no parents. A naive Bayes model with Gaussian $P(X_i|C)$ is equivalent to a mixture of Gaussians with diagonal covariance matrices. While mixtures of Gaussians are widely used for density estimation in continuous domains, naive Bayes models have seen very little similar use in discrete and mixed domains. However, they have some notable advantages for this purpose. In particular, they allow for very efficient inference of marginal and conditional distributions. To see this, let X be the set of query variables, Z be the remaining variables, and k be the number of mixture components (i.e., the number of values of C). We can compute the marginal distribution of X by summing out C and Z:

$$P(X = x) = \sum_{c=1}^{k} \sum_{z} P(C = c, X = x, Z = z)$$

$$= \sum_{c=1}^{k} \sum_{z} P(c) \prod_{i=1}^{|X|} P(x_i|c) \prod_{j=1}^{|Z|} P(z_j|c)$$

$$= \sum_{c=1}^{k} P(c) \prod_{i=1}^{|X|} P(x_i|c) \prod_{j=1}^{|Z|} \sum_{z_j} P(z_j|c)$$

$$= \sum_{c=1}^{k} P(c) \prod_{i=1}^{|X|} P(x_i|c)$$

where the last equality holds because, for all j, $\sum_{z_j} P(z_j|c) = 1$. Thus the non-query variables Z can simply be ignored when computing P(X = x), and the time required to compute P(X = x) is O(|X|k), independent of |Z|. This contrasts with Bayesian network inference, which is worst-case exponential in |Z|. Similar considerations apply to conditional probabilities, which can be computed efficiently as ratios of marginal probabilities: $P(X = x|Y = y) = P(X = x, Y = y)/(Y = y)$. A slightly richer model than naive Bayes which still allows for efficient inference is the mixture of trees, where, in each cluster, each variable can have one other parent in addition to C. The basic graph of Bayesian network is presented in **Figure 1** and the graph of QWS dataset for naïve Bayes are depicted in **Figure 3**.

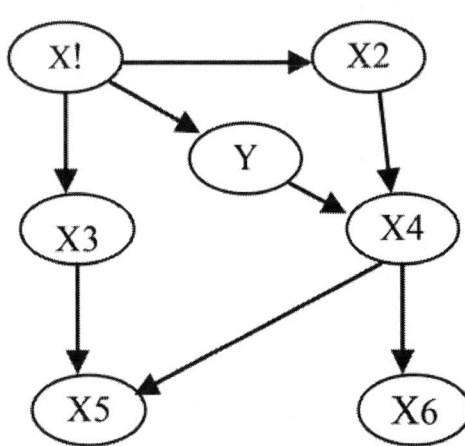

**Figure 1**. The Bayesian network (S, P).

**Figure 2.** A Markov blanket DAG for Y.

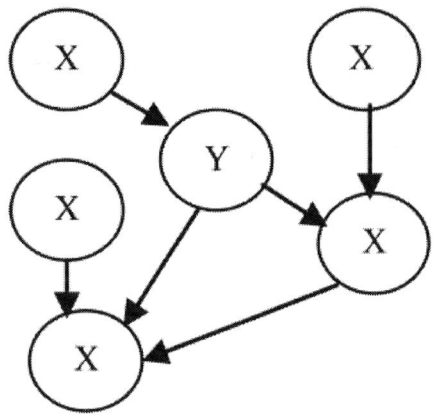

## 3.2. Markov Blanket

The Markov condition implies that the joint distribution P can be factorized as a product of conditional probabilities, by specifying the distribution of each node conditional on its parents [20]. In particular, for a given DAG S, the joint probability distribution for X can be written as

$$P(X) = \prod_{i=1}^{n} P_i(X_i | Pa_i) \quad (3)$$

where $Pa_i$ denotes the set of parents of $X_i$ in S; this is called a Markov factorization of P according to S. The set of distributions represented by S is the set of distributions that satisfy the Markov condition for S. If P is faithful to the graph S, then given a Bayesian Network (S, P), there is a unique Markov blanket for Y consisting of PaY, the set of parents of Y, chY, the set of children of Y, and Pa chY, the set of parents of children of Y.

For example, consider the two DAGs in Figures 1 and 2, above. The factorization of P entailed by the Bayesian Network (S, P) is

$$P(Y, X_1, \cdots, X_6) = P(Y|X_1) \cdot P(X_4|X_2, Y)$$
$$\cdot P(X_5|X_3, X_4, Y) \cdot P(X_2|X_1) \cdot P(X_3|X_1) \quad (4)$$
$$\cdot P(X_6|X_4) \cdot P(X_1)$$

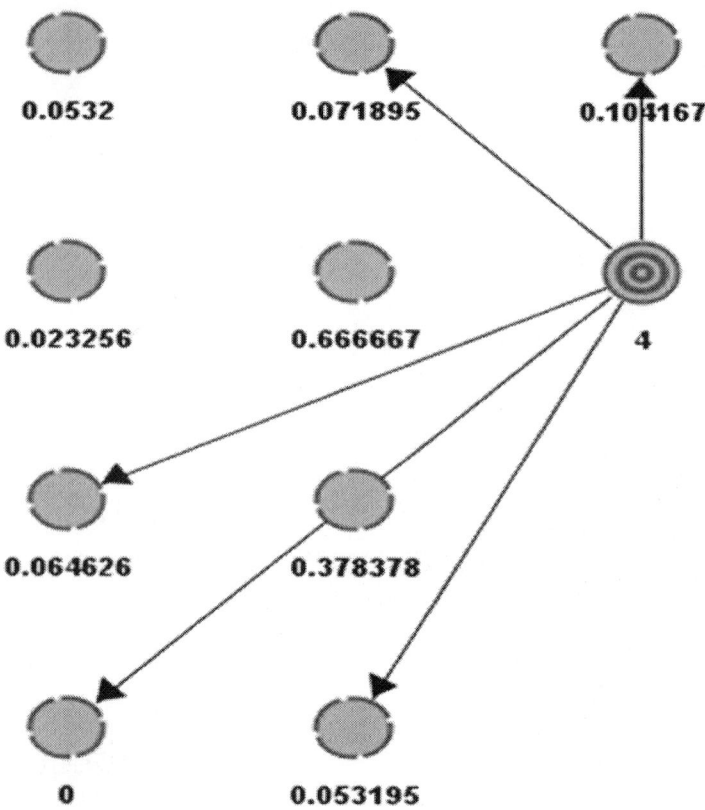

**Figure 3.** QWS dataset for Naïve Bayes bayesian network.

The factorization of the conditional probability p(Y |X$_1$, ···, X$_6$) entailed by the Markov blanket for Y corresponds to the product of those (local) factors in equation (2) that contain the term Y.

$$P(Y|X_1,\cdots,X_6) = C_0 \cdot P(Y|X_1)$$
$$\cdot P(X_4|X_2,Y) \cdot P(X_5|X_3,X_4,Y) \quad (5)$$

The graph of QWS dataset is depicted in **Figure 5** for Markov blanket of Bayesian networks.

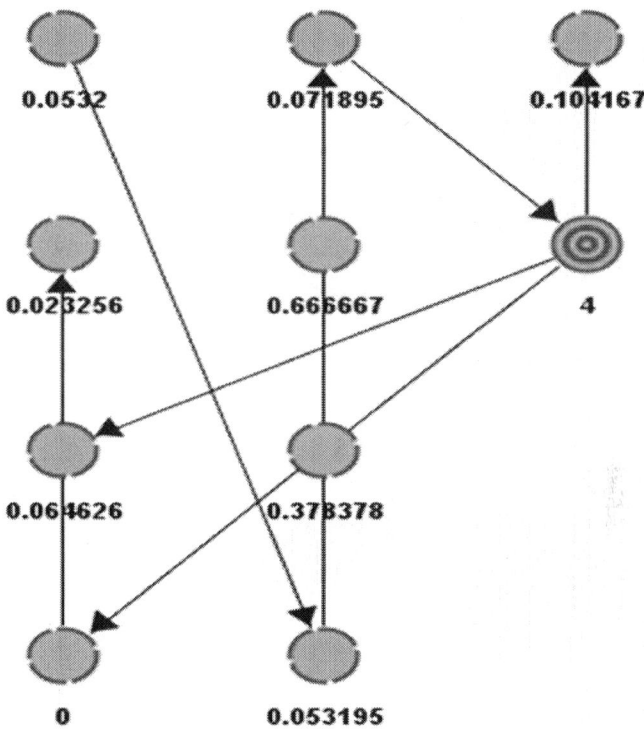

**Figure 4.** QWS dataset for Tabu search.

## 3.3. Tabu Search

A Heuristic is an algorithm or procedure that provides a shortcut to solve complex decision problems. Heuristics are used when you have limited time and/or information to make a decision. For example, some optimization problems, such as the travelling salesman problem, may take far too long to compute an optimal solution. A good heuristic is fast and able find a solution that is no more than a few percentage points worse than the optimal solution. Heuristics lead to good decisions most of the time, but not always. There have been several Meta-heuristic applications recently in Machine Learning, Evolutionary Algorithms, and Fuzzy Logic problems. Tabu Search is a meta-heuristic strategy that is able to guide traditional local search methods to escape the trap of local optimality with the assistance of

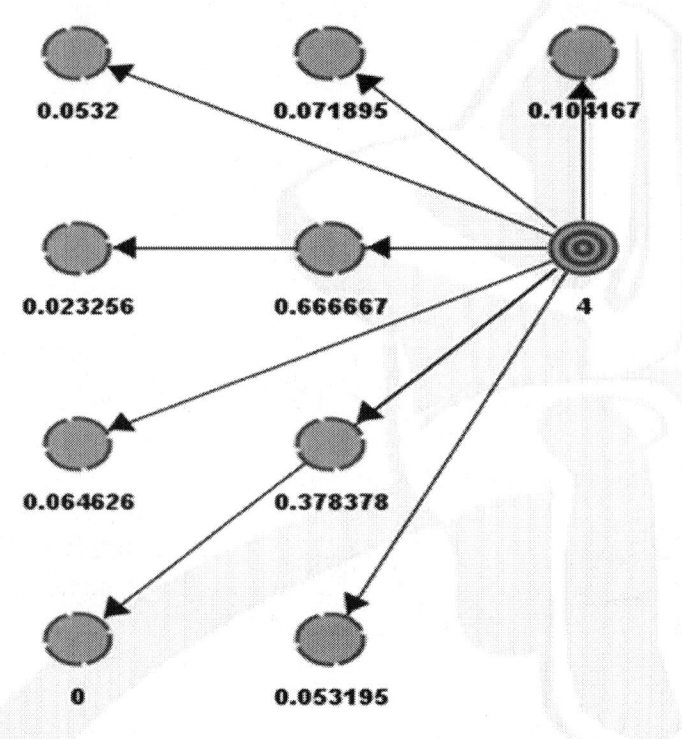

**Figure 5.** QWS dataset for Markov blanket bayesian network.

adaptive memory [21]. Its strategic use of memory and responsive exploration is based on selected concepts that cut across the fields of artificial intelligence and operations research.

Tabu list (TL) is given by

$$TL = \{s^{-1} : s = s_i,\ i > k - t,\} \tag{6}$$

where k is the iteration index and $s^{-1}$ is the inverse of the move s; i.e., $s^{-1}(s(x)) = x$. In words, TL is the set of those moves that would undo one of those moves in the t most recent iterations. t is called the Tabu tenure.

The use of TL provides the "constrained search" element of the approach, and hence the solution generated depends critically on the composition of TL and the way it is updated. Tabu search makes no reference to the condition of local optimality, except implicitly where a local optimum improves on the best solution. A best move, rather than an improving move is chosen at each step. This strategy is embedded in the OPTIMUM function. Tabu search analysis of QWS attribute is depicted in **Figure 4**.

## 4. RESULTS AND DISCUSSIONS

We employed Naïve Bayes, Markov blanket and Tabu search techniques to classify web services. We note that the average accuracy of Naïve Bayes classifier is 85.62%, followed by Tabu search of 82.45% and Markov blanket of 81.36% presented in **Table 3**. In this context, we employed Back propagation trained neural network to find the importance of different attributes in web services. We found that we found that WRRF plays a vital role for classifying the web services. We excluded the WSRF from dataset and experimented. We observe that the average accuracy of Naïve Bayes is 75.01% and Marcov Blanket is 65.48% and Tabu serach is 71.38% presented in **Table 4**. As Bayesian network is a very good classifier to classify classification type of problems. In this context, the result obtained by Bayesian classifier is not superior to the results obtained by [22] to classify the accuracy of web services. As Bayesian network has not been applied to classify web services, we employed this approach in the present study.

**Table 3**. Accuracies of bayesian network without removing wsrf.

| Classifiers | Accuracy (%) |
|---|---|
| Naïve Bayes | 85.62 |
| Markov Blanket | 81.36 |
| Tabu Search | 82.45 |

**Table 4**. Accuracies of bayesian nework after removing wsrf.

| Classifiers | Accuracy (%) |
|---|---|
| Naïve Bayes | 75.01 |
| Markov Blanket | 65.48 |
| Tabu Search | 71.38 |

## 5. CONCLUSION

We presented Naïve Bayes, Markov blanket and Tabu search to classify web services. We observed that Naïve Bays approach predicted better accuracy than Markov blanket and Tabu search. Secondly, Bayesian belief network is employed first time to classify web services in the present study. Future directions include more exploration of Markov blanket approach for rule generation and quality of attributes to decide the classification of web services.

## 6. ACKNOWLEDGEMENTS

We are grateful to Dr. E. Al-Masri and Dr. Q. H. Mahmoud for providing us the dataset related to the web services classification.

## REFERENCES

1. L. Z. Zeng, B. Benatallah, M. Dumas, J. Z. Kalagnanam and Q. Z. Sheng, "Web Engineering: Quality Driven Web Service Composition," Proceedings of the 12th International Conference on World Wide Web, Budapest, 2003, pp. 411-421.
2. A. Tsalgatidou and T. Pilioura, "An Overview of Standards and Related Technology in Web Services," Distributed and Parallel Database, Vol. 12, No. 2-3, 2002, pp. 135-162.
3. P. Domingos and M. Pazzani, "On the Optimality of the Simple Bayesian Classifier under Zero-One Loss", Machine Learning, Vol. 29, No. 2-3, 1997, pp. 103-130.
4. P. Cheeseman and J. Stutz, "Bayesian Classification (Auto Class): Theory and Results," Advances in Knowledge Discovery and Data Mining, Vol. 180, 1996, pp. 153-180.
5. D. Margaritis, "Learning Bayesian Network Model Structure," Technical Report CMU-CS-03-153, 2003.
6. I. Tsamardinos, C. F. Aliferis and A. Statnikov, "LargeScale Feature Selection Using Markov Blanket Induction for the Prediction of Protein-Drug Binding," Technical Report DSL-02-08, Vanderbilt University, Nashville, 2002.

7. I. Tsamardinos, C. Aliferis, and A. Statnikov, "Algorithms for Largescale Local Causal Discovery in the Presence of Small Sample or Large Causal Neighborhood," Technical Report DSL-02-08, Vanderbilt University, Nashville, 2002.
8. D. M. Chickering, "Learning Equivalence Classes of Bayesian-Network Structures," Journal of Machine Learning Research, Vol. 2, 2002, pp. 507-554.
9. S. Vinoski, "Service Discovery 101," Proceedings of Internet Computing Conference on IEEE, Vol. 7, No.1, 2003, pp. 69-71.
10. G. V. Bochmann, B. Kerherve, H. Lutffiyya, M.-V. M. Salem and H. Ye, "Introducing QoS to Electronic Commerce Applications," Proceedings of the 2nd International Symposium on Electronic Commerce (ISEC), HongKong, 26-28 April, 2001 pp. 138-147.
11. A. Mani and A. Nagarajan, "Understanding Quality of Services for Web Services," 2002. http://www-106.ibm.com/developerworks/library/ws-quality
12. S. Ran, "A Model for Web Services Discovery with QoS," ACM SIGecom Exchange, Vol. 4, No. 1, 2003.
13. M. Conti, E. Gregori and F. Panzieri, "Load Distribution among Replicated Web Services A: QoS Based Approach," Second Workshop on Internet Server Performance, ACM Press, 1999.
14. O. Ardaiz, F. Freitag and L. Navarro, "Improving Service Time of Web Clients Using Server Redirection," ACM SIGMETRICS Performances Evaluation Review, Vol. 29, No. 2, 2001, pp. 39-44.
15. S. Kalepu, S. Krishnaswamy and S. W. Loke, "Verity: A QoS Metric for Selecting Web Services and Providers," Proceedings of the 4th International Conference on Web Mining System Engineering Workshops (WISEW'03), 2004, pp. 131-139.
16. S. Araban and L. Sterling, "Measuring Quality of Service for Contract Aware Web-Services," First Australian Work- shop on Engineering Service-Oriented Systems (AWESOS 2004), Melbourne, 29 January 2004, pp. 116-127.
17. QWS Datasets. http://www.uogue/ph.ca/~qmahmoud/qws/index.html
18. E. AL-Masri and Q. H. Mahmoud, "Investing Web Services on the World Wide Web," The 17th International Conference on World Wide Web, Beijing, 21-25 April 2008, pp. 795-804.

19. E. AL-Masri and Q. H. Mahmoud, "QoS Based Discovery and Ranking of Web Services," IEEE 16th International Conference on Computer Communications and Networks (ICCCN), Honolulu, 13-16 August 2007, pp. 529-534.
20. J. Pearl, "Probabilistic Reasoning in Intelligent Systems: Networks of Plausible Inference," Morgan Kaufmann, 1988.
21. F. Glover and M. Laguna, "Tabu Search," Kluwer Academic Publishers, Amsterdam, 1997.
22. R. Mohanty, V. Ravi and M. R. Patra, "Web Services Classification Using Intelligent Techniques," Expert Systems with Applications, Vol. 37, No. 7, 2010, pp. 5484- 5490.

# CHAPTER 5

## LANE-LEVEL ROAD INFORMATION MINING FROM VEHICLE GPS TRAJECTORIES BASED ON NAÏVE BAYESIAN CLASSIFICATION

Luliang Tang [1], Xue Yang [1,*], Zihan Kan [1] and Qingquan Li [1,2]

[1]State Key Laboratory of Information Engineering in Surveying, Mapping, and Remote Sensing, Wuhan University, Wuhan 430079, China

[2]Shenzhen Key Laboratory of Spatial Smart Sensing and Services, College of Civil Engineering, Shenzhen University, Shenzhen 518060, China

## ABSTRACT

In this paper, we propose a novel approach for mining lane-level road network information from low-precision vehicle GPS trajectories (MLIT), which includes the number and turn rules of traffic lanes based on naïve Bayesian classification. First, the proposed method (MLIT) uses an adaptive density optimization method to remove outliers from the raw GPS trajectories based on their space-time distribution and density clustering. Second, MLIT acquires the number of lanes in two steps. The first step establishes a naïve Bayesian classifier according to the trace features of the road plane and road profiles and the real number of lanes, as found in the training samples. The second step confirms the number of lanes using test samples in reference to the naïve Bayesian classifier using the known trace features of test sample. Third, MLIT infers the turn rules of each lane through tracking GPS trajectories. Experiments were conducted using the GPS trajectories of taxis in Wuhan, China. Compared with human-interpreted results, the automatically generated lane-level road network information was demonstrated to be of higher quality in terms of displaying detailed road networks with the number of lanes and turn rules of each lane.

## KEYWORDS

GPS trajectories; adaptive density optimization method; naïve Bayesian classifier; lane-level information; big data

## 1. INTRODUCTION

Accurate lane-based road network data for navigation, such as lane location, lane changes, and turn information, is crucial for ensuring reliable and safe driving, especially for intelligent transportation systems (ITS) such as advanced driver assistance systems and autonomous driving. In addition, lane geometry information, especially the number of lanes, can also be important for inferring the type of road and for estimating traffic flow capacity. Conventional road networks, which are extracted by digitization, mobile mapping vehicles, or aerial photographs, are based on road centerlines [1,2]. Lane-level information (such as number of lanes and turning in the intersection) is usually acquired from high-definition video/images, laser point clouds, or DGPS/INS trajectories [3,4,5,6,7]. Mining such detailed information is time consuming and labor intensive [8].

Increasingly, public vehicles and personal navigation assistants are equipped with single-frequency global position system (GPS) trackers or loggers that monitor the user locations at regular intervals [9,10]. The quality of tracking data is often low due to the geometrical effects of urban canyons on the accuracy of GPS ranging, thereby causing tracking points to deviate from the original roads. However, large volumes of data can be inexpensively acquired using GPS tools. This new type of geospatial resource contains abundant information regarding road networks, traffic conditions, points of interest, and driving behaviors [11,12,13]. Extracting high-quality road maps from low-quality tracking data is a hot topic [14,15,16,17]. As compared to the existing approaches for generating lane-based network information, a geospatial approach takes full advantage of information generated by spatial and temporal tracking data, thus enabling a user to establish or update road networks (e.g., road-level network and lane-level network) in relation to traffic rules.

This study proposes an approach (MLIT: mining lane-level road network information from vehicle GPS trajectories) to automatically generating lane-level road information including number of lanes and intersection turns from low-precision vehicle GPS trajectories gathered from thousands of taxis in a city. To reduce the impact of low accuracy and other vagaries in taxi trajectories, we take steps to offset the uncertainty of lane-based road network extraction. First, trajectories gathered in off-peak times comprise the experimental data [18,19]. Second, segmentation treatment is adopted during optimization and lane information extraction to avoid the significant impact of large vibrations on the spread of trajectories due to the changes in traffic features such as uncertain traffic flows and lane additions at different positions on the same road. Therefore, we define the trajectory segment section (TSS) as the basic unit for trajectory optimization and lane information extraction. Specifically, according to the trace features of the road plane and road profile, and the real number of lanes found in the training samples, we construct a naïve Bayesian classifier to infer the number of lanes from the test samples based on these trace features. The turn rules of each lane, including going straight, or making a left, right or U-turn, are extracted by tracking trajectories in relation to the rate of reckless driving. The contributions of this paper are as follows:

1. We propose a new method, the adaptive density optimization method, for vehicle GPS trajectory optimization based on the density clustering method and the spatial distribution of tracking points. Outliers mixed in the raw data are removed automatically using adaptive density optimization method.
2. We explore a novel way to infer lane-level information from low-precision spatiotemporal vehicle GPS trajectories (MLIT).
3. We detect turn rules of each lane by tracking vehicle trajectories in relation to the rate of reckless driving.

The remainder of this article is organized as follows. In Section 2, related studies on outlier removal and extracting lane information from low-precision GPS trajectories are reviewed. In Section 3, we fully describe the proposed method for detecting traffic lane information. In Section 4, a series of experiments on Wuhan datasets demonstrate the advantages and effectiveness of the proposed method. In Section 5, the conclusions and directions of future research are discussed.

## 2. RELATED WORK

Spatial trajectories are never perfectly accurate, due to sensor noise and other factors. Sometimes the error is acceptable, such as when using GPS to identify the city where a person is located. In other situations, various methods to remove noise and decrease the error in the measurements are applied to trajectory data. Specifically, trajectories gathered by public vehicles include massive amounts of real-time and low-cost information (e.g., road network, point of interests, driving behaviors). This information also contains many outliers due to the limitations of the collection equipment, the environment, and purpose. The main categories for outlier removal include filtering algorithms to smooth the noise, map-matching methods to change the original coordinates of tracking points to fit them to the existing road network, and clustering methods to remove outliers.

Filtering is important when the trajectory data is particularly noisy, or when it is necessary to derive other quantities like speed or direction from these data. In addition, filtering is suitable for trajectories with high sampling frequency only. Lee in [20] gave a detailed introduction on how to implement a filtering algorithm, including the Kalman filter and particle filter. Another pre-processing step uses map-matching methods to match the trajectories to the road network. Sotiris Brakatsoulas [21] presented three map-matching algorithms that focus especially on the trajectory nature of the data rather than simply on the current position, as in traditional map-matching techniques. It is important to note that map-matching methods apply only to road-level information extraction from GPS data because each tracking point is matched to the centerline of the carriageway, and its original location is changed. In contrast to these methods, some researchers have used clustering for outlier removal. For example, Jing Wang [15] proposed using kernel density methods to remove outliers among raw GPS trajectories. Chen [7] sorted all the data points in ascending order according to their distances from the median and then chose 95% of the sorted data points as the experimental data. Compared with filtering techniques and map-matching methods, clustering methods do not change the position of tracking points, and are less susceptible to sampling intervals, making them most suitable for outlier removal when dealing with a large volume of low-quality GPS trajectories at a low sample frequency and affected by urban canyons. However, in our case, the experimental data is collected in an urban area and their position accuracy and sampling rate are about 10–15 m and 20 s respectively. The kernel density estimation with a fixed bandwidth for outlier removal is not suitable for a complex road network in an urban area. In addition, the outlier's proportion of low-precision GPS data far exceeds 5%. Therefore, we are motivated to use an adaptive density estimation method to remove outliers.

After data pre-processing, automatic road network refinement from GPS trajectories becomes possible. OpenStreetMap employs user-contributed GPS trajectories to create free digital maps that are open for editing by registered users [22]. Likewise, WikiMapia, Google Maps, and other map applications let users update maps. There has also been work on completely automated methods to infer road maps fromlow-quality GPS trajectories. Those methods include matching GPS traces to prototypical shapes [23], using incremental methods to process GPS traces and generate road maps [24,25], and applying clustering methods or artificial algorithms to extract road networks from GPS traces [26,27,28]. According to the references [23,24,25,26,27,28], the existing methods can generate and update road-level maps from low-quality vehicle GPS trajectories, while latter studies addressing detailed road network generation have gradually shifted down to lane-level road network information.

Lane-level information extraction from vehicle trajectories starts with high-precision differential GPS data and concludes with a refinement of an existing map, including locating lanes and number of lanes [29,30]. This process involves smoothing and filtering the GPS data, matching it to an existing map, spline fitting the road centerline, clustering to find lanes, and refinement of the intersection geometry. Uduwaragoda [31] proposed using high-precision vehicle trajectories collected by vehicles equipped with INS/GPS-enabled mobile phones to generate lane-level road maps based on kernel density estimation. However, the methods proposed in [29,30,31] are based on the assumption that GPS traces from different lanes are separated well. For low-precision GPS data, this assumption is seriously violated, and therefore a kind of probabilistic method with prior knowledge is used to extract lane structure from a mass of low-precision GPS trajectories. Moreover, previous study [31] did not consider detailed lane extraction across large areas or regions, nor did it focus on turn extraction. Our study contributes to the existing research not only by generating lane-level road network information, including the number and turn rules of traffic lanes from preprocessing low-precision vehicle GPS trajectories, but also empirically evaluating the validity of the results for large areas and regions.

## 3. LANE-LEVEL ROAD NETWORK INFORMATION EXTRACTION FROM VEHICLE GPS TRAJECTORIES

In this section, we present the lane-level road network information (e.g., number of lanes and turns of each lane) extraction method (MLIT) from low-precision GPS trajectories. The architecture is shown in Figure 1. The processing of MLIT is described as follows:

First, outliers in each TSS (trajectory segment section) are removed automatically with the adaptive density optimization algorithm.

Second, a naïve Bayesian classifier is constructed by analyzing the trace feature ($x^{(1)}$) of the road plane and trace feature ($x^{(2)}$) on the road profile and the real number of lanes in the training samples.

Third, according to the naïve Bayesian classification, the number of lanes of test sample is inferred by reference to naïve Bayesian classifier with known $x^{(1)}$ and $x^{(2)}$ of test sample.

Finally, the turn rules of each lane are inferred by tracking GPS trajectories.

### 3.1. Vehicle GPS Trajectory Optimization

Density-based clustering methods are relatively suitable for spatial trajectories because they can reveal clusters of arbitrary shapes and can filter out noise [32]. A vehicle GPS trajectory does not always overlap with the actual path of a vehicle due to GPS positioning error. A statistical analysis of the locations where GPS points are dense suggests a high probability that a road is present whereas a low density of points indicates that vehicles either deviated too far from the road or moved along a small branch road with few trajectories. Based on these considerations, points from multiple trajectories with low density are considered as outliers. In this section, we adopt an adaptive density optimization method to eliminate outliers. In addition, segmentation treatment optimization avoids significant large vibrations on

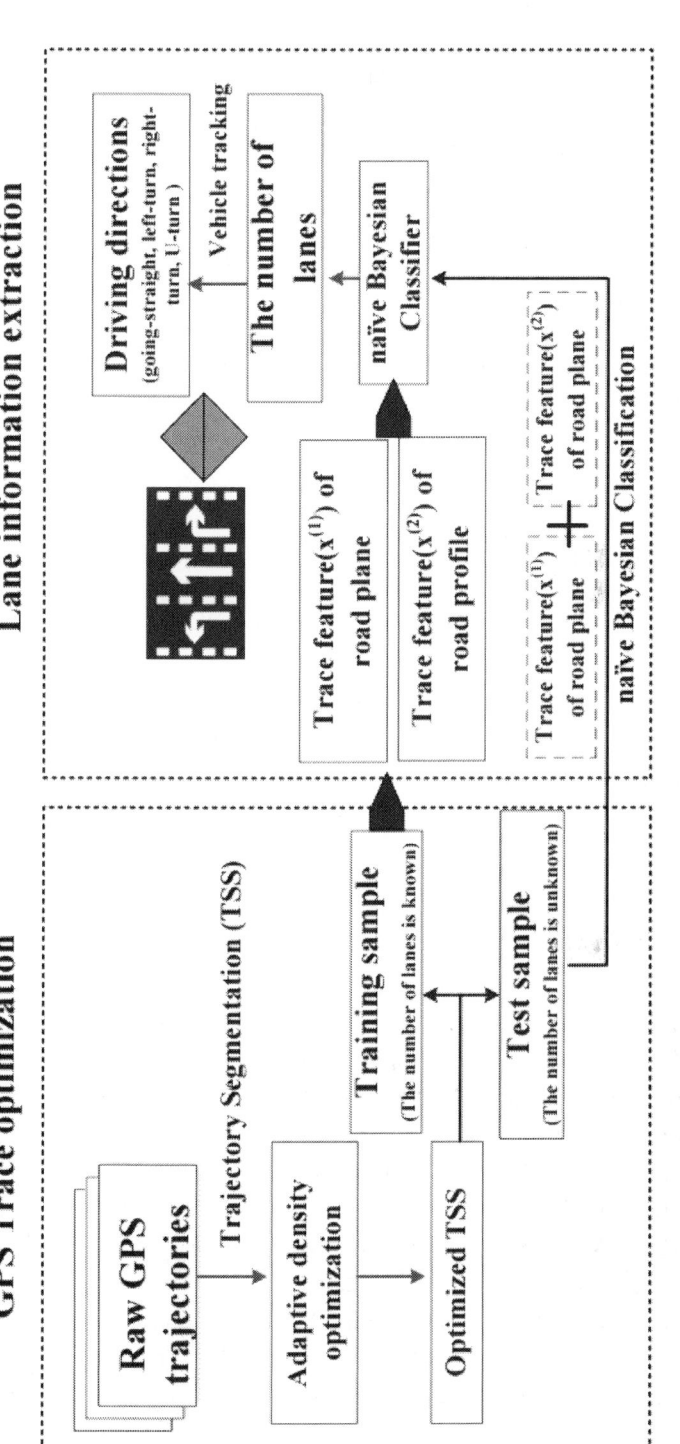

**Figure 1.** Lane information extraction architecture.

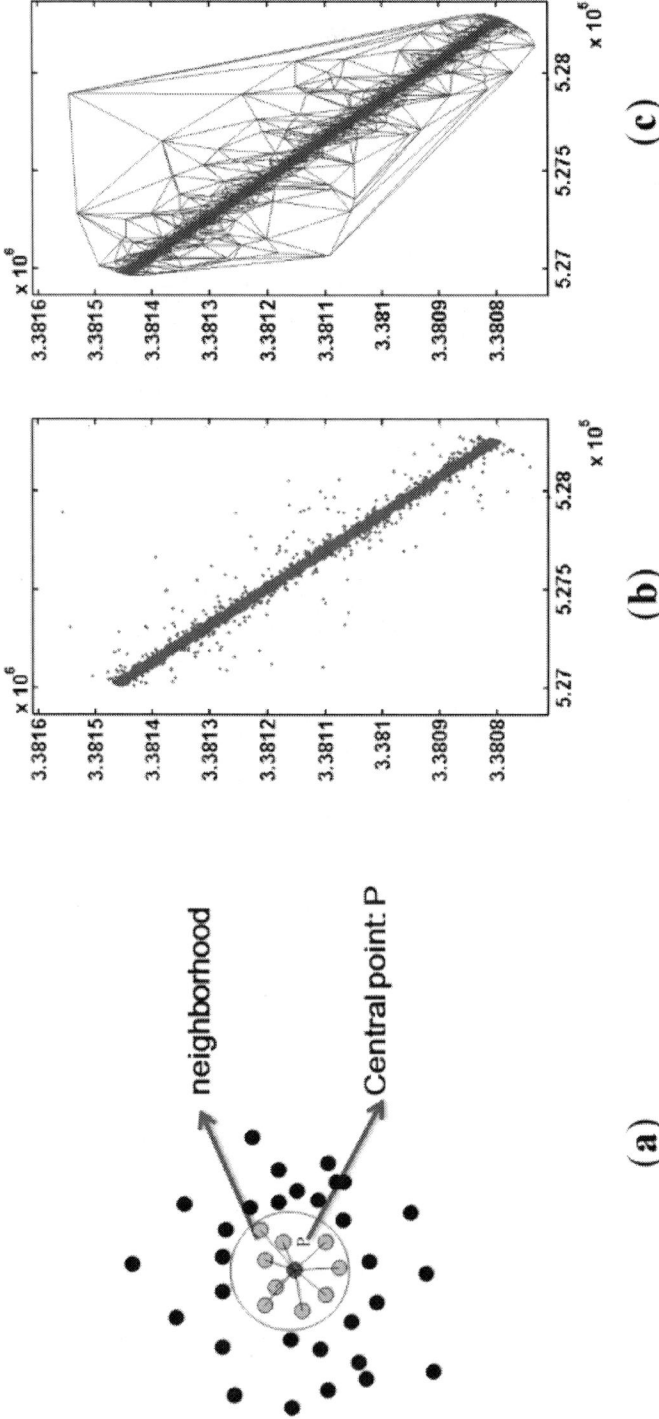

**Figure 2.** Trajectories optimization. (**a**) shows the distribution of tracking points, *p* is the central point and the circle in (a) is the neighborhood of *p*; (**b**) indicates a subset of tracking points and be denoted as *A*; (**c**) is the Delaunay triangulation of *A*.

the spread of trajectories due to the changes in traffic features such as uncertain traffic flows and lane additions at different positions on the same road.

### 3.1.1. Adaptive Density Optimization Method

A tracking point P is defined as high-density point if the points in the neighborhood of tracking point P display a high clustering degree (Figure 2a). The density of points can be represented as a significance and tested to identify significant high-density point clusters. The null distribution [33,34] of any point dataset can be defined as:

$$P(N(A)=k) = \frac{\lambda^k \left(|A|^k\right)}{k!} e^{-\lambda|A|} \qquad (1)$$

where $N(A)$ is the number of points in any subset denoted by $A$ (e.g., Figure 2b), $k = 1, 2, \ldots, N(A)$, $|A|$ is the area of subset $A$, $\lambda$ is the intensity of the null distribution, and can be estimated as:

$$\lambda = \frac{N(A)}{|A|} \qquad (2)$$

Therefore, the significance of the aggregation of points in neighborhood can be calculated as:

$$P(X \geq n_i) = 1 - \sum_{m=0}^{n_i-1} \frac{\left(|Nei|^m\right) e^{-\lambda|Nei|\lambda^m}}{m!} \qquad (3)$$

where $x$ denotes any point in the subset $A$, $P(x \geq n_i)$ is the significance of $x$, $n_i$ is the number of the points in the neighborhood of $x$, $|Nei|$ is the area of neighborhood of $x$, $r$ is the radius of neighborhood of $x$.

$$|Nei| = \pi r^2 \qquad (4)$$

In order to simplify the calculation, the radius $r$ is used to indicate $|Nei|$. Thus Equation (3) can be simplified as:

$$P(X \geq n_i) = 1 - \sum_{m=0}^{n_i-1} \frac{\left(|r|^m\right) e^{-\lambda |r| \lambda^m}}{m!} \tag{5}$$

For different point datasets, the radius $r$ is adaptive, depending on the spatial distribution of points. Thus, based on a Delaunay triangulation, the radius of neighborhood is statistically defined as:

$$r = meanDE + variationDE \tag{6}$$

where *meanDE* is the mean length of all edges of the Delaunay triangulation, and *variationDE* is the standard deviation of the length of all edges in the Delaunay triangulation (see Figure 2c). The area of $A$ is computed as:

$$|A| = \sum_{i=1}^{M} AT_i \tag{7}$$

where $M$ is the number of triangles in the Delaunay triangulation (see Figure 2c), $AT_i$ is the area of triangle $i$. Each tracking point density is computed using formulas 1–7. Then the proposed method compares the density to significance $\eta$ (usually set $\eta = 0.05$ or $\eta = 0.01$), and $x$ is defined as an actual tracking point if its significance is less than η; otherwise the point $x$ is defined as an outlier and removed from the dataset.

### 3.1.2. Optimization

In order to avoid significant large vibrations on the spread of trajectories due to the changes of traffic features, such as the uncertainty of traffic flow and adding of lanes at different positions on the same road, we define a trajectory segment section (TSS) as the basic unit for trajectory optimization and lane information extraction. Each TSS is obtained by dividing the trajectory segment (TS) with a fixed length, $h$, as shown in Figure 3. The fixed length $h$ for dividing TS depends largely on road construction rules of

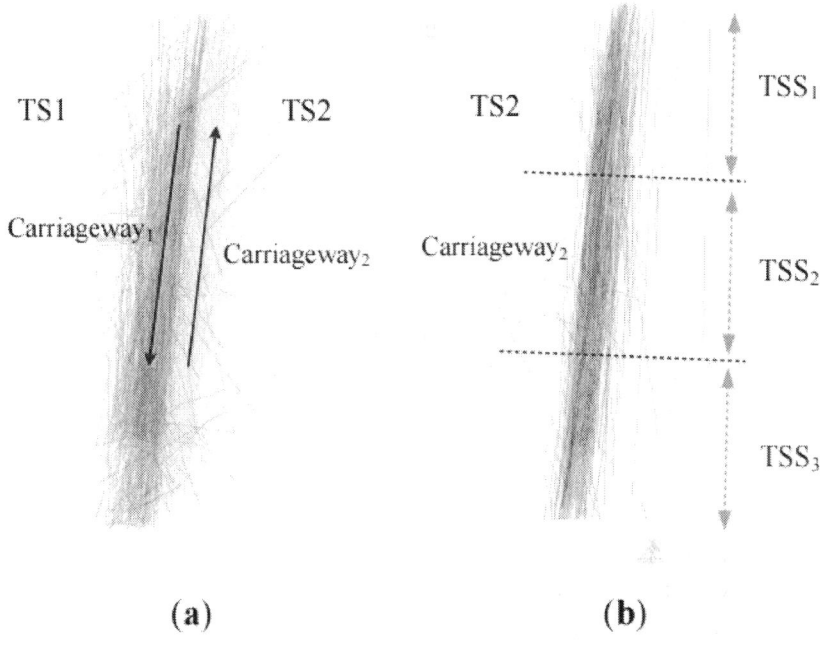

**Figure 3.** TS and TSS. (**a**) is the description of TS and (**b**) is the depiction of TSS.

a city. For example, adding a lane on a road generally occurs when the road is within 50 m of an intersection; elsewhere, the length of a lane added on a road as a parking area for buses or taxis often is greater than 10 m. Therefore, the fixed length should be less than 50 m and greater than 10 m for a better result when extracting the number of lanes. The details for acquiring TS and TSS were recommended in [35].

We assume that any $TS_i$ contains several TSSes denoted as ($TSS_1$, $TSS_2$, ..., $TSS_n$). The $TSS_i$ has the point set *SubB* and the region of Delaunay triangulation network. In Figure 4a, the black point is the part of $TS_i$ and the red points belong to $TSS_i$. In Figure 4b, the red line shows the Delaunay triangulation for $TSS_i$. The radius of neighborhood can be computed based on Equation (6) and the area of *SubB* is calculated based on Equation (7). In this way, we can avoid the limitations caused by using the same radius of neighborhood to optimize trajectories in different density regions. Moreover, the accuracy when extracting the number of lanes improves because the added lanes on a road can be detected through segment treatment.

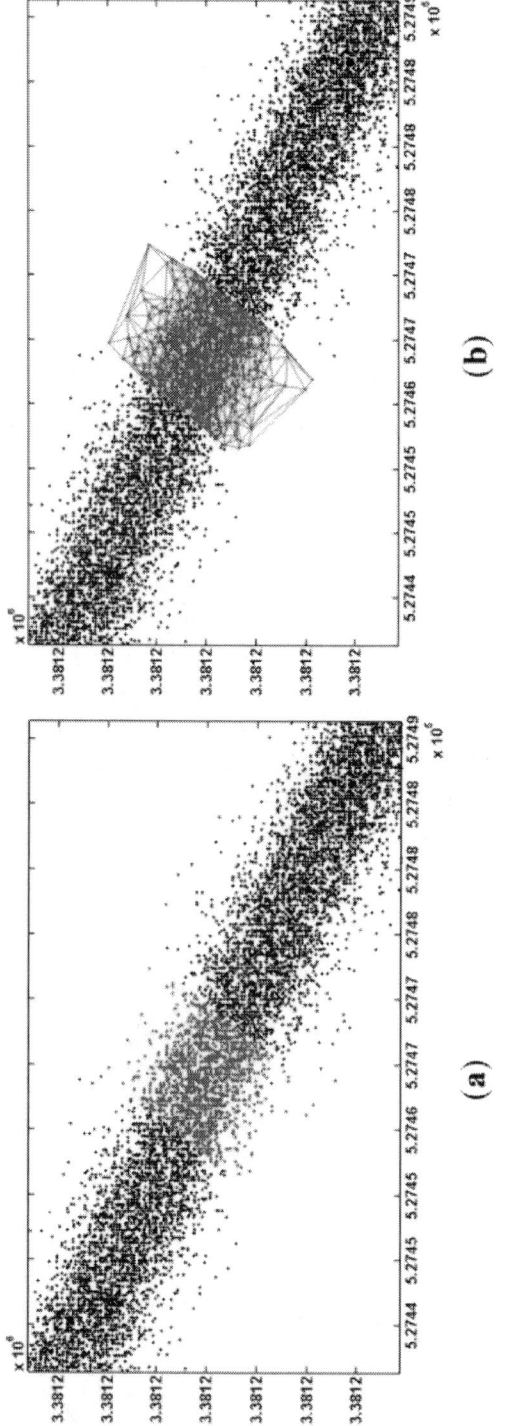

**Figure 4.** Trajectories optimization way. (**a**) is the tracking points of $TS_i$; (**b**) shows the Delaunay triangulation network of $TSS_i$.

## 3.2. Lane Number Extraction Based on Naïve Bayesian Classification

Although low-precision vehicle trajectories with low-sampling frequency have the advantage of low costs and a short gathering period, they are still limited by gathering accuracy and frequency even when optimized. Those limitations mean that the final lane-level clustering results might differ from the actual structure of lanes. Therefore, based on the trace feature of road plane ($x^{(1)}$) and trace feature of road profile ($x^{(2)}$), *a priori* knowledge of training samples was introduced as a constraint into the lane number extraction, and the naïve Bayesian classification of lane number extraction was proposed. The implementations are introduced in detail in an upcoming section.

### 3.2.1. The Basic Method

For training samples $T = \{TSS_1(x_1, y_1), TSS_2(x_2, y_2), \ldots, TSS_N(x_N, y_N)\}$, sample $TSS_i$ associated with trace feature set $x_i$ and category label set $y_i$, $x_i = (x_i^{(1)}, x_i^{(2)})$ and $y_i = (c_1, c_2, \ldots, c_K)$, where $x_i^{(1)}$ and $x_i^{(2)}$ are trace features of road plane and road profile, $c_i$ is number of lanes in a road. For instance $x_i^{(j)}$, $x_i^{(j)} \in x_i$, has several possible values such as $x_i^{(j)} \in \{a_{j1}, a_{j2}, \ldots, a_{j1}, \ldots, a_{jsj}\}$, $a_{jl}$ is the possible value for $x_i^{(j)}$. Then the prior probability $P(Y = c_k)$ and conditional probability $P(x^{(j)} = a_{jl} | Y = c_k)$ can be calculated as:

$$P(Y = c_k) = \frac{\sum_{i=1}^{N} I(y_i = c_k)}{N} \tag{8}$$

$$P(X^{(j)} = a_{jl} | Y = c_k) = \frac{\sum_{i=1}^{N} I(x_i^{(j)} = a_{jl}, y_i = c_k)}{\sum_{i=1}^{N} I(y_i = c_k)} \tag{9}$$

where $I$ is the indicated function, $I_y(c) = 1$ if $c \in y$ and 0 otherwise, $i = 1, 2, \ldots, N$, $j = 1, 2$, $l = 1, 2, \ldots, sj$, $k = 1, 2, \ldots, K$.

Given test instance $x = (x^{(1)}, x^{(2)})^T$, $x^{(1)}$ and $x^{(2)}$ are the trace feature of test instance $x$, the posterior probability is defined as:

$$P(Y=c_k|X=x) = \frac{P(Y=c_k)\prod_j P(X^{(j)}=x^{(j)}|Y=c_k)}{\sum_k P(Y=c_k)\prod_j P(X^{(j)}|Y=c_k)} \quad (10)$$

Based on the above notations and Bayesian rule, the lane number of test instance $x$ is determined as:

$$y = \arg\max_{c_k} P(Y=c_k)\prod_{j=1}^{2} P(X^{(j)}=x^{(j)}|Y=c_k) \quad (11)$$

where $y$ is the number of lanes in test instance $x$, and $y \in (c_1, c_2, ..., c_K)$.

### 3.2.2. Naïve Bayesian Classifier

The naïve Bayesian classification determines the number of lanes in a TSS, and a naïve Bayesian classifier is thus available. The key to constructing a naïve Bayesian classifier is to acquire trace feature set $x_i$ and category set $y_i$ from the training samples $T$. In this study, the naïve Bayesian classifier acts as *a priori* knowledge for the number of lane extraction from test samples. The categorization of roads by number of lanes in the training samples must include all lane types in a city.

Trace feature set $x_i$ of training samples includes two aspects: trace feature of road plane ($x^{(1)}$) and trace feature of road profile ($x^{(2)}$). Millions of vehicles travel around each road in the city and gather a massive amount of tracking points that include information such as location, speed, direction and vehicle ID. A large number of tracking points cover a whole road and their coverage width will gradually become stable with an increasing number of trajectories. Road width also closely relates to the number of lanes. Thus, a trace feature of a road plane ($x^{(1)}$) is the trajectories strip width (TSW) and indicates the real width of road with certain accuracy after outlier removal. Additionally, trajectories distributed on the same lane always cluster together. Thus, a trace feature of road profile ($x^{(2)}$) is a cluster of a number of trajectories along the road cross-section and indicates the number of lanes to some extent.

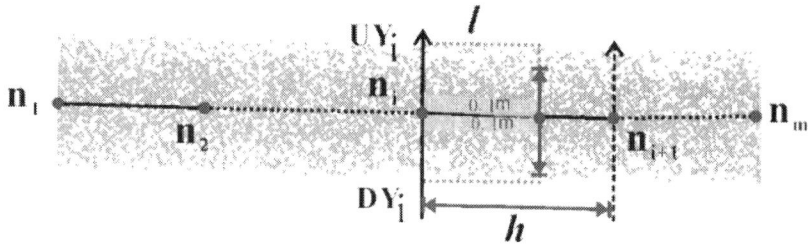

**Figure 5.** The detection of trajectories strip width.

(1) Trace Feature of Road Plane Extraction

In this paper, we propose the adaptive width detection method to detect the TSW (Figure5).

As shown in Figure 5, the total length of the trajectory segment $TS_i$ that starts at an intersection and ends at another intersection is $L$. If the fixed length is set as $h$, we can successively get $m$ ($m = L/h$) sections as the trajectory segment section (TSS), and the diverging points of each TSS are recorded as $\{n_1, n_2, ..., n_m\}$. The direction of the trajectory segment $TS_i$ and the diverging points are respectively set as the horizontal axis and origins of the width detection coordinate system. $UY_j$ and $DY_j$ are the positive and negative directions of the longitudinal axis, respectively. The details of the algorithm are as follows:

---

/*Initialization*/
Coordinate origin: $n_1$;
horizontal axis: the direction of the current *TSS*;
longitudinal axis: $UY_i = 0$; $DY_i = 0$;

Sliding window: length = $l$; width = $w$; proportion = 0;
/*Assignment*/
for each $TSS_i$, do
repeat

Moving the sliding window along the positive direction and negative direction of the longitudinal axis and accumulating the Proportion (Proportion = current points number in sliding window/all points in the current *TSS*)

until proportion = 100%

set $Dw_i = \sum$ (maximum $|UY_j| + |$ maximum $|DY_j|)/(h/l); j = 1,2,..., (h/l)$.
set Coordinate origin changed to $n_{i+1}$; $UY_{i+1} = 0$; $DY_{i+1} = 0$; $i = 1,2,...,m$.
**end for**

---

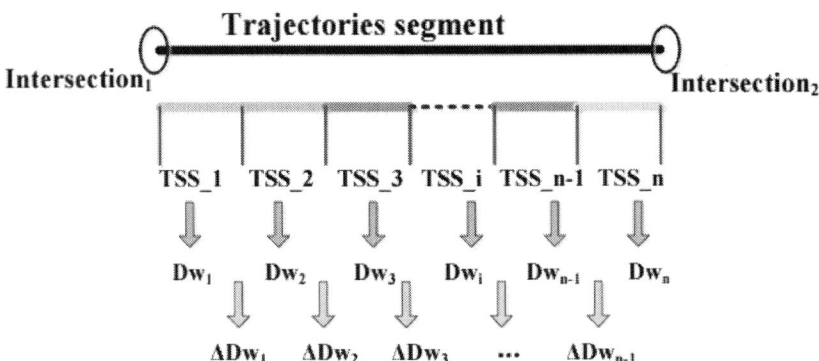

**Figure 6.** Trajectories strip width analysis.

The results for the TSS width ($Dw_1, Dw_2, ..., Dw_m$) are obtained as shown in Figure 6, where $\Delta Dw_1, ..., \Delta Dw_{n-1}$ are the differences of each TSS width.

In most cases, the value $\Delta Dw_i$ between two adjacent TSSes will stay within one lane width $a$ due to position accuracy of GPS data or from added lanes. However, some abnormal results still exist due to the effects of temporary parking areas, bus stops, dense lanes appearing near intersections, etc., which make $\Delta Dw_i$ abnormally larger than $a$. Thus, the nearest measurement result of TSS replaces abnormal results considering that the width along the road is always in a relatively steady state. We will explain this approach in more detail.

**Step 1**, compute the difference of each TSS width between $Dw_i$ and $Dw_{i+1}$, that is, $\Delta Dw_i = Dw_i - Dw_{i+1}$ ($i = 1, 2, ..., n - 1$);

**Step 2**, compare each $\Delta Dw_i$ with $a$, $Dw_i$ is replaced by $Dw_{i+1}$ when $\Delta Dw_i$ is larger than $a$ and $Dw_i$ is larger than $Dw_{i+1}$ ($i = 1, 2, ..., n - 1$);

**Step 3**, do step1 and step 2 repeatedly until all abnormal values are optimized.

The results for TSS width difference are shown in Figure 7, the blue line indicates the raw results from $\Delta Dw_i$, the red points are the abnormal values, and the green line shows the result of optimized $\Delta Dw_i$.

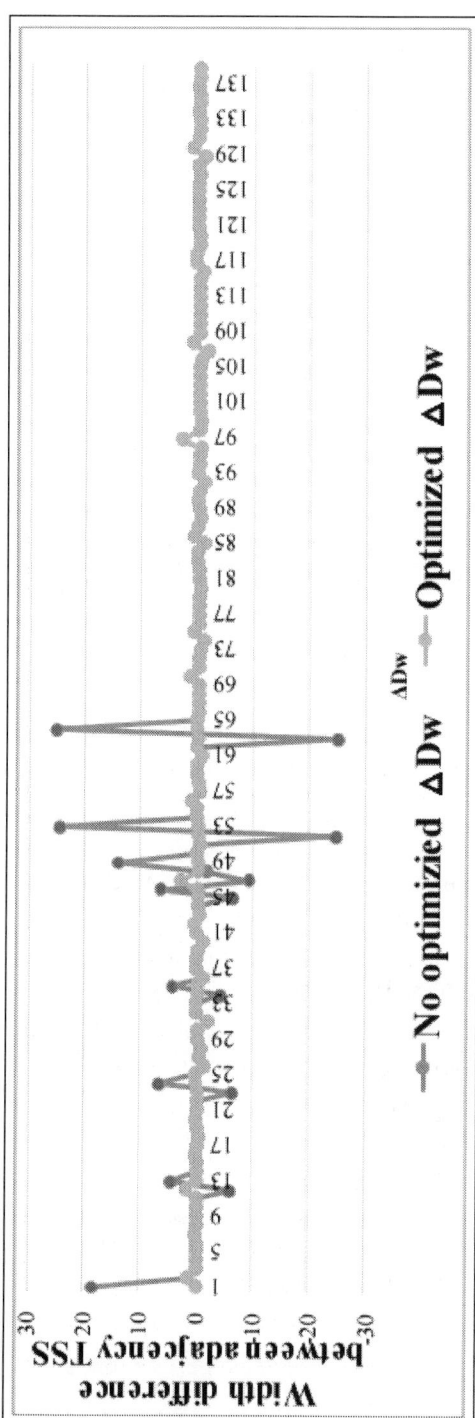

**Figure 7.** Width detection results proprecessing.

## (2) Trace Features of Road Profile Extraction

According to methods described previously, in the naïve Bayesian classifier, the trace feature of road profile ($x^{(2)}$) is the number of clustered trajectories. The authors of [7,29,30,31] proposed using cluster methods to detect lane structure from high-precision GPS trajectories with high sampling frequency such as partition clustering, hierarchical clustering and statistical clustering. According to reference [7] statistical clustering is more suitable for lane structure detection from ordinary GPS data at an accuracy within 4 m. Therefore, for acquiring cluster number of trajectories in road profiles, we fit a constrained Gaussian mixture model (CGMM) to perpendicular cross-sections of the traces across the road, based on the assumption that GPS trajectories will tend to cluster near the center of each lane with some spread due to GPS noise and other vagaries. The CGMM can be defined as:

$$p(x) = \left( \sum_{j=1}^{ln} \omega_j \frac{1}{\sqrt{2\pi\sigma^2}} \exp\left(-\frac{(x-\mu_j)^2}{2\sigma^2}\right) \right) \tag{12}$$

where $ln$ represents the number of Gaussian components, and each component corresponds to each lane, while $w_1, \ldots, w_{ln}$ are the weights of each component, corresponding to the relative traffic volume in each lane and $\sum_{j=1}^{ln} w_j = 1$. The $\mu_1, \ldots,$ is the means of the trajectories for each component and equals to the centerline of each lane; $\sigma$ is the standard variance of the trajectories for each component and set same value because the width of lane of adjacency lane usually is the same. The number of components for a CGMM is equal to cluster number of trajectories of a TSS and determined by the structural risk model (structural risk minimization, SRM). To estimate $w_i$, $u_i$ and $\sigma$ for a set of $ln$'s and then select the $ln$ that minimizes the structural risk model. The method to calculate and extract the number of clusters can be obtained according to [7].

In addition, a trajectory is a line described by a series of points. Each point has a gathering time and spatial location, such as $Trace_i = \{p_1, p_2, \ldots, p_n\}$, $p_i = (x_i, y_i, t_i, direction_i, speed_i, state_i)$ ($i = 1, 2, \ldots, n$), where $(x_i, y_i)$ is the spatial location, $t_i$ is the gathering time, and $direction_i$ and $speed_i$ give the motion status of a moving object. At the same time, $state_i$ is attributed information of a moving object such as the ID number of a moving object, as shown in Figure 8a. However, this description is not appropriate for ana-

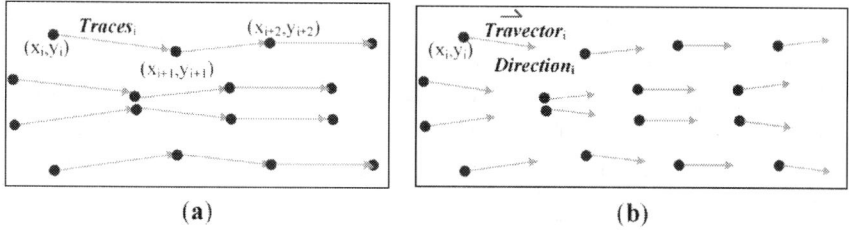

**Figure 8.** Trajectories and trajectory vector description. (**a**) shows the trajectory vector constructed according to traditional style; (**b**) is the trajectory vector proposed in this paper.

lyzing the longitudinal density distribution of low-quality trajectories because the sampling intervals of this kind of data range from ten seconds to one minute, making the distance between any two adjacent points too large to retain enough information. Thus, we replace common trajectories with trajectory vectors, and obtain a number of trajectories clusters by detecting the longitudinal density distribution of those trajectory vectors.

For each trajectory vector, the tracking point is regarded as the start point, the direction of tracking point as vector direction, and the speed as the vector mold. The tracking point is denoted as $P_i$ ($x_i$, $y_i$, $t_i$, $direction_i$, $speed_i$, $state_i$), and its trajectory vector is described as ***Travector**_i* = ($x_i$, $y_i$) and |***Travector**_i*| = $speed_i \times \Delta t$, $\Delta t$ = 1 s ($\Delta t$ does not affect the final outcome for GMM computation, this paper set it as 1 s), as shown in Figure 8b.

To facilitate CGMM computation, we rotate the axes so that the X axis is made parallel to the average direction vector. Here, the rotation matrix in Equation (13) is used. The angle $\xi$ can be obtained according to [35].

$$\begin{bmatrix} x' \\ y' \end{bmatrix} = \begin{bmatrix} \cos\xi & \sin\xi \\ -\sin\xi & \cos\xi \end{bmatrix} \begin{bmatrix} x \\ y \end{bmatrix} \qquad (13)$$

The longitudinal density distribution of trajectories is acquired by projecting each trajectory vector to the vertical axis ($y'$). The projecting ordinate of each trace vector is set as the sampling point and replaces the intersection points of trajectories and sampling lines perpendicular to the road centerline.

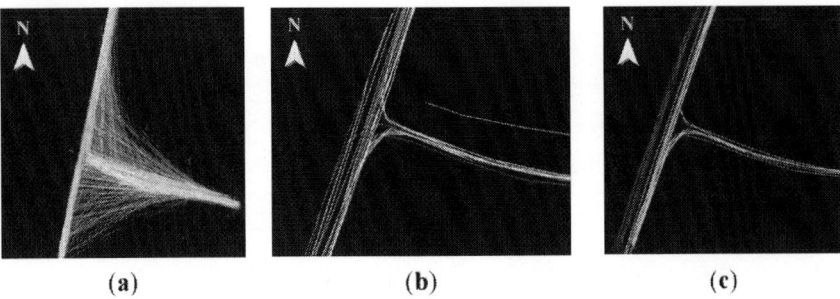

**Figure 9.** Trajectory tracking. **(a)** indicates the trajectories with 40 s sampling interval; **(b)** shows the trajectories with 20 s sampling interval; **(c)** descripts the different driving directions of vehicles.

### 3.3. The Detection of Turn Rules of Each Lane

GPS trajectories are a sequence of GPS points with the time interval between any consecutive GPS points not exceeding a certain threshold $\Delta T$ ($\Delta T$ is the sampling interval.), as shown in Figure 9. We detect the turn rules of each lane by tracking GPS trajectories. Figure 9 illustrates the trajectories of vehicles. Figure 9a shows trajectories at a 40 s sampling interval. Figure 9b indicates the trajectories at a 20 s sampling interval. Figure 9c shows different driving directions of vehicles and the sample rate of trajectories is 20 s, where the red lines represent the turn rules of each vehicle from north to south, the green lines denote the south to north direction, the blue and yellow lines indicate vehicle right turns.

Through recording the trajectory segments, the change of trajectory direction is replaced by the change of the trajectory segments' direction. Figure 10 indicates the trajectory belonging to segments $TS_{001}$ and $TS_{004}$. The change of direction between $TS_{001}$ and $TS_{004}$ is computed as $\Delta\theta = \theta_2 - \theta_1$; these ($\theta_1, \theta_2$) are directions of $TS_{001}$, $TS_{004}$ obtained by [34]. The turn rules of the lane traversed by the trajectory are "left Turn," "right turn," "going straight" and "U-turn," if change of direction satisfies the conditions: ($\Delta\theta < 0°$ & $\Delta\theta \approx -90°$), ($\Delta\theta > 0°$ & $\Delta\theta \approx 90°$), ($\Delta\theta \approx 0°$) and ($\Delta\theta > 0°$ & $\Delta\theta \approx 180°$), respectively. The turn rules of each lane are further determined by Equation (14).

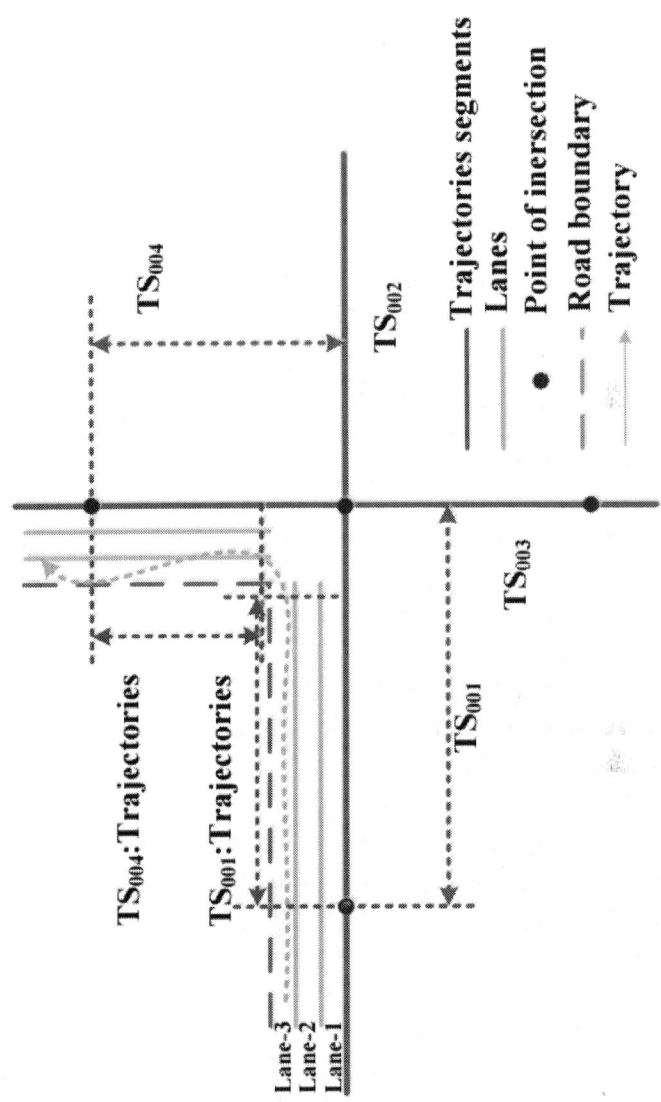

**Figure 10.** Intersection turns: left turn, right turn and U-turn detection.

$$f_i = \frac{value_i}{\sum_{i=1}^{4} value_i} \quad (i = 1, 2, 3, 4) \tag{14}$$

That $value_i$ is the number of GPS trajectories belonging to $group_i$ ($i$ = 1, 2, 3, 4 indicates "left turn," "right-turn," "going straight" and "U-turn," respectively) on the lane. The final rate of $group_i$ on the lane is denoted as $f_i$. The turning of the lane is $group_i$ if $f_i$ is far beyond a predefined rate of reckless driving.

## 4. EXPERIMENTS AND RESULTS

Our test GPS data came from thousands of taxis driving in Wuhan city, as shown in Figure 11a. The sampling frequency ranges from 10 s to 40 s, while the positioning accuracy ranges from 10 m to 15 m. Each taxi was recorded for an average of 14 days, and we collected in total about 200 billion GPS points, as shown in Figure 11b. We obtained about 2000 TS, and 300,000 TSS when the fixed length $h$ was set as 10 m. The number of trajectory vectors in each TSS ranges from 100 to 1000.

### 4.1. Trajectory Optimization

Outliers mixed in raw GPS trajectories were eliminated using the adaptive density optimization method. The trajectory strip width (TSW) of each TSS was obtained and optimized using our proposed adaptive width detection method. The length $l$ and width $w$ of the sliding window were set as 10 m and 0.1 m, respectively; the significance $\eta$ was set as 0.05, as recommended by reference [33]; and the width of lane $a$ for optimized trajectories was set as 3.75 m according to the road construction standards in China. Figure 12 shows the results of trajectory optimization, where red points and black points represent the valid data and outliers, respectively.

An evaluation for trajectory optimization results was done by comparing with the correlation between TSW before and after optimization and the actual road width, as shown in Figure 13. The test data (about 400 TSS) were randomly selected from the 300,000 TSS, the trajectory strip width

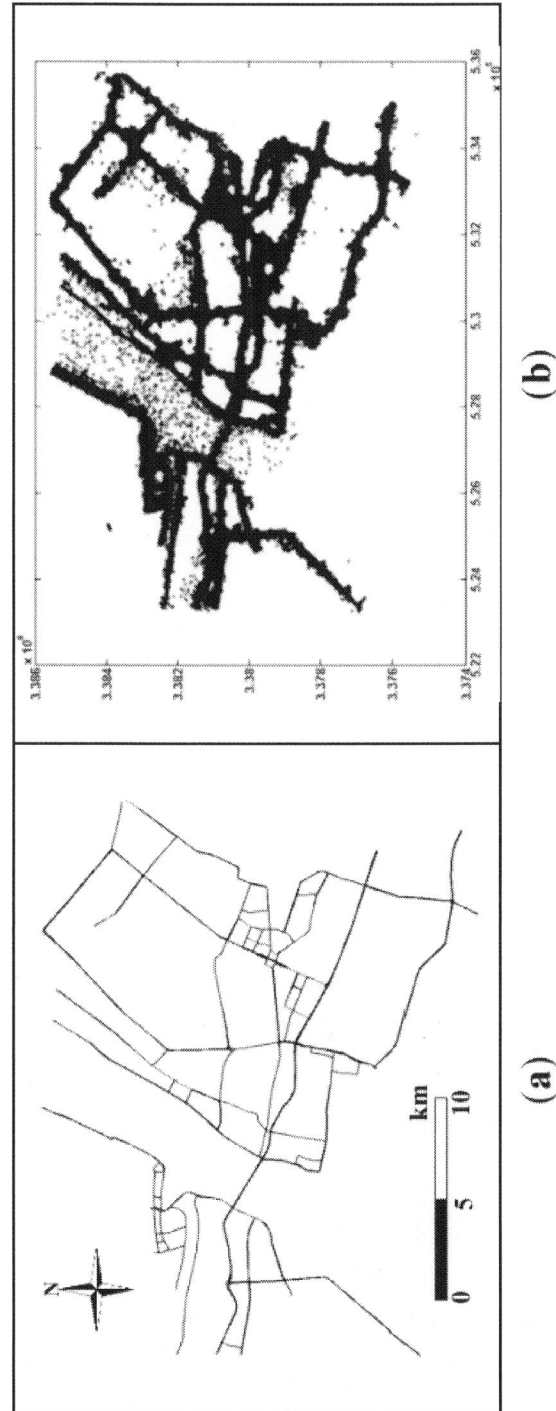

**Figure 11.** Experimental data. (**a**) is the road network of the experimental area; (**b**) shows the raw trajectories collected by taxis.

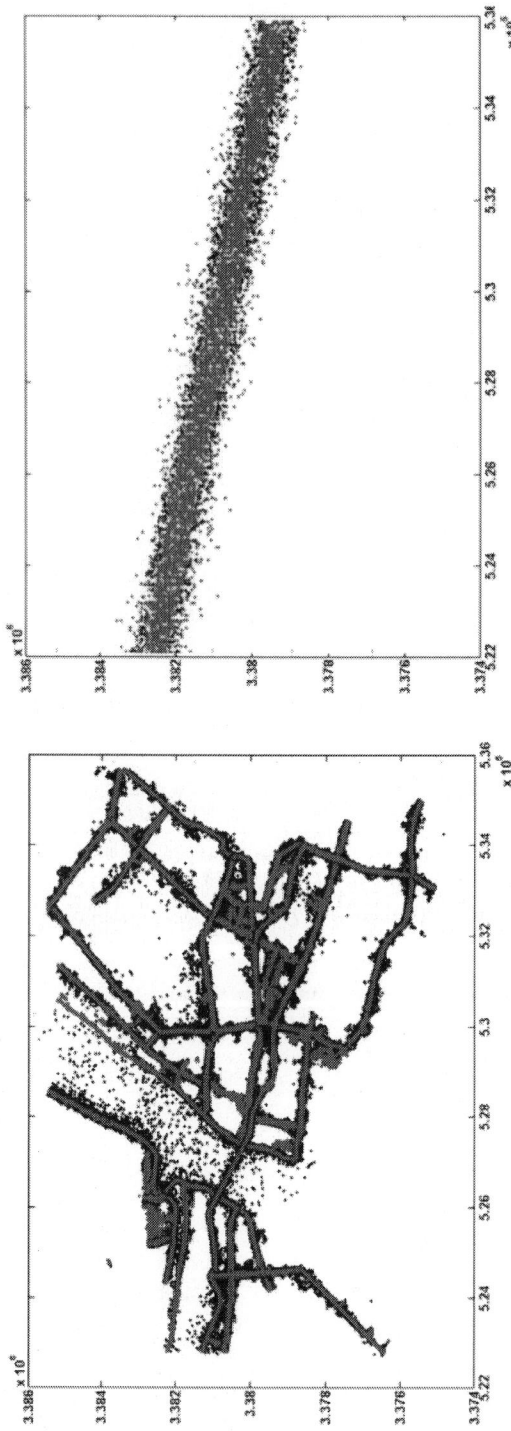

**Figure 12.** Optimization results. The result of all experimental data (**left**); the magnification of one segment (**right**).

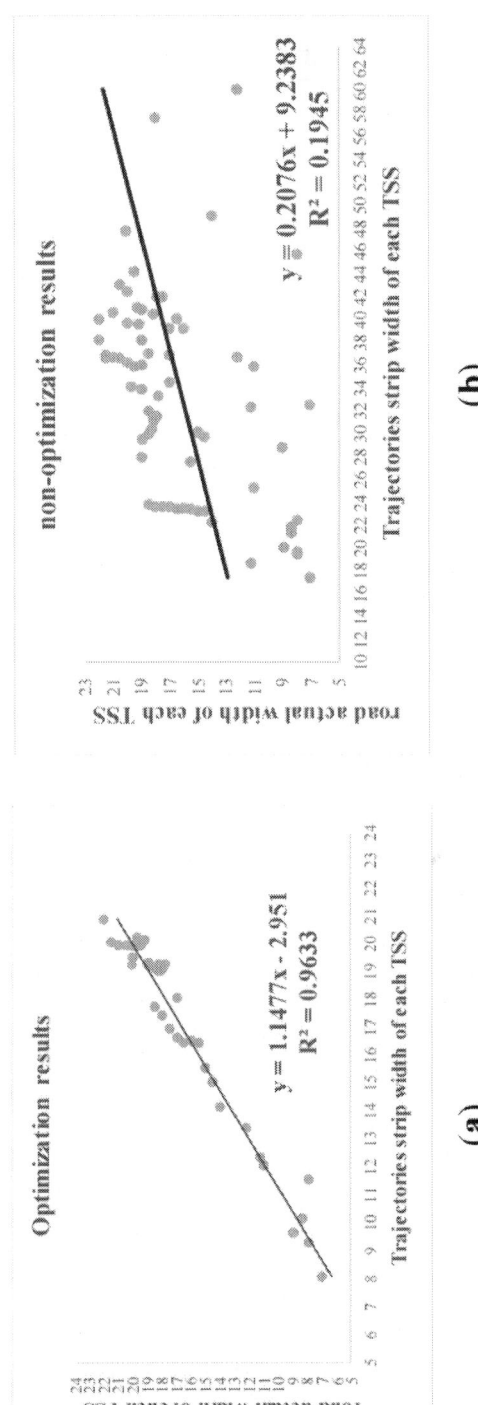

**Figure 13.** Optimization results evaluation: (**a**) shows the evaluation results of optimized trajectories; (**b**) indicates the evaluation results of non-optimized trajectories.

(TSW) before and after optimization was acquired using adaptive width detection algorithm, and the actual width of the road was obtained by field measurement. Based on statistics, the strong correlation ($R^2 = 0.9633$) between TSW and actual width of road of each optimized TSS illustrates that TSW of optimized TSS is very close to the actual road width. Comparing with the non-optimization results (Figure 13b), the optimization results indicate that the proposed optimization method performs well.

## 4.2. The Construction of Naïve Bayesian Classifier

A selection of TSSes (about 7,650 TSSes) found on various types of lanes in the experimental area were used as training samples, and the other 11,350 TSSes were designated as test samples. The real number of lanes in the training sample was extracted by observing a corresponding remote sensing image. By analyzing the relation between the real number of lanes, the trajectory strip width (TSW) and number of clusters in the training sample, we established a naïve Bayesian classifier, as listed in Table 1.

Table 1 indicates the categories of the number of lanes in the experimental area including two-lane, three-lane, four-lane and five-lane roads. Each type contains three values, such as the value of TSW ($x^{(1)}$), the number of trajectories clusters ($x^{(2)}$) and the real number of lanes ($y$) for the training sample.

## 4.3. Lane Information Extraction

Given a test instance TSS $(13.8m, 3)^T$, the lane number is calculated as:

$P(y = 2) = 278/765; P(y = 3) = 220/765; P(y = 4) = 189/765; P(y = 5) = 78/765;$
$P(x^{(1)} = 13.8 \mid y = 2)=0/278; P(x^{(2)} = 3 \mid y = 2) = 98/278;$
$P(x^{(1)} = 13.8 \mid y = 3) = 220/220; P(x^{(2)} = 3 \mid y = 3) = 185/220;$
$P(x^{(1)} = 13.8 \mid y = 4) = 189/189; P(x^{(2)} = 3 \mid y = 4) = 34/189;$
$P(x^{(1)} = 13.8 \mid y = 5) = 0/78; P(x^{(2)} = 3 \mid y = 5)=5/78;$
$P(y = 2) * P(x^{(1)} = 13.8 \mid y = 2) * P(x^{(2)} = 3 \mid y = 2) = 0;$
$P(y = 3) * P(x^{(1)} = 13.8 \mid y = 3) * P(x^{(2)} = 3 \mid y = 3) = 0.242;$
$P(y = 4) * P(x^{(1)} = 13.8 \mid y = 4) * P(x^{(2)} = 3 \mid y = 4) = 0.044;$
$P(y = 5) * P(x^{(1)} = 13.8 \mid y = 5) * P(x^{(2)} = 3 \mid y = 5) = 0.$

**Table 1.** Naïve Bayesian classifier.

| Training Sample (ID) | Trace Feature: $x^{(1)}$/m | Trace Feature: $x^{(2)}$ | Category Label Set: $y$ |
|---|---|---|---|
| 1 | 7.9–12.2 | 2 | 2 |
| 2 | 7.9–12.2 | 3 | 2 |
| ... | ... | ... | ... |
| 2,780 | 7.9–12.2 | 2 | 2 |
| 2,781 | 10.2–19.8 | 3 | 3 |
| 2,782 | 10.2–19.8 | 3 | 3 |
| ... | ... | ... | ... |
| 4,980 | 10.2–19.8 | 4 | 3 |
| 4,981 | 13.2–20.8 | 4 | 4 |
| 4,982 | 13.2–20.8 | 4 | 4 |
| ... | ... | ... | ... |
| 6,870 | 13.2–20.8 | 3 | 4 |
| 6,871 | 17.6–25.8 | 4 | 5 |
| 6,872 | 17.6–25.8 | 5 | 5 |
| ... | ... | ... | ... |
| 7,650 | 17.6–25.8 | 5 | 5 |

Thus, according to Equation (11), the number of TSS lanes were 3. Specifically, the road centerline of TSS was acquired according to [26], then we inferred the lane boundary based on the number of lanes of TSS with the width of lane $a$. Turn rules of each lane is determined according to Equation 14. At the same time, according to the road construction standard in China, the lane width $a$ is set to 3.75 m, and the predefined rate of reckless driving for turn information extraction is set as 5%. Table 2 indicates the other results of lane information detection for each TSS, including number of lanes and driving directions of each lane.

In Table 2, most results show the stability and the validity of lane information extraction using a naïve Bayesian classifier, but a few mistakes still occurred. For example, numbers of lanes such as $TSS_{001}$ of $TS_{002}$, and $TSS_{042}$ of $TS_{003}$ were misclassified. We use arrow-shaped indications to represent the driving directions of each lane, where arrow-shaped indication ↑ indicates that vehicle drivers go straight, and multi-headed arrows ↱ show that vehicle drivers can travel in straight direction or turn left at an intersection, as shown in Table 2. At the same time, the accuracy of turn rules of lane detection depends largely on the results of the number of lanes. In Table 2, the turn rules from lanes in the test samples also get a misclassification because of an incorrect estimate of the number of lanes.

**Table 2.** Lane information identification.

| TS | TSS | $x^{(1)}$/m | $x^{(2)}$ | The Number of Lanes (Detections) | The Number of Lanes (True Value) | Driving Direction (Detections) | Driving Direction (True Value) |
|---|---|---|---|---|---|---|---|
| TS$_{001}$ | TSS$_{001}$ | 10.1 | 2 | 2 | 2 | ↑↑ | ↑↑ |
| | TSS$_{002}$ | 9.9 | 2 | 2 | 2 | ↑↑ | ↑↑ |
| | ... | ... | ... | ... | ... | ... | ... |
| TS$_{002}$ | TSS$_{001}$ | 14.1 | 4 | 4 | 3 | ↑↑↑↱ | ↑↑↱ |
| | TSS$_{002}$ | 14.2 | 4 | 4 | 3 | ↑↑↱ | ↑↑↱ |
| | ... | ... | ... | ... | ... | ... | ... |
| | TSS$_{016}$ | 15.4 | 3 | 3 | 3 | ↑↑↱ | ↑↑↱ |
| | ... | ... | ... | ... | ... | ... | ... |
| TS$_{003}$ | TSS$_{001}$ | 19.2 | 4 | 4 | 4 | ↑↑↑↱ | ↑↑↑↱ |
| | TSS$_{002}$ | 20.3 | 4 | 4 | 4 | ↑↑↑↱ | ↑↑↑↱ |
| | TSS$_{003}$ | 20.3 | 3 | 3 | 4 | ↑↑↑↱ | ↑↑↑↱ |
| | ... | ... | ... | ... | ... | ... | ... |
| | TSS$_{042}$ | 20.3 | 5 | 5 | 3 | ↑↑↑↱↱ | ↑↑↱ |

## 4.4. Quantitative Evaluation

### 4.4.1. Quantitative Evaluation for Number of Lane Identification

To evaluate the performance of our proposed method for detecting lane information, we compared test samples for the number of lanes extraction to those manually marked. Table 3 shows quantitative values for precision, recall, and f-score in the proposed method (MLIT) and the methods of [29,30,31]. This comparison demonstrates that MLIT has better precision, recall, and f-score in lane number extraction than those of [29,30,31]. Meanwhile, the result of this comparison shows that MLIT is more suitable for low-precision GPS trajectories with low-sampling frequency than other methods of [29,30,31] that detects lane structure directly from raw trajectories, but does not consider the prior knowledge during lane number identification. At the same time, experimental results also authenticate that low-precision GPS trajectories from different lanes are not separated well. In addition, Table 3 shows that the proposed method extracts the lane numbers with an overall accuracy of 83.72%; however, there is also a 16.68% chance of incorrectly identifying the number of lanes. The reasons are as follows.

**Table 3.** Quantitative evaluation and comparison.

| Methods | Precision | Recall | F-Score |
|---|---|---|---|
| MLIT | 83.72% | 83.35% | 84.03% |
| Kernel density estimation [31] | 78.48% | 79.05% | 78.76% |
| Hierarchical agglomerative clustering [30] | 63.23% | 62.7% | 62.96% |
| K-means clustering [29] | 62.54% | 61.51% | 62.02% |

First, MLIT has a difficulty in lane number identification because a small number of complex intersections (e.g., the incorrect results in Table 2) have different traffic flows between adjacent lanes caused by traffic lights, driving restrictions, and other traffic characteristics.

Second, MLIT cannot distinguish trajectories from roads on and below viaducts, since the experimental data has no elevation information. The lane information for overlapping roads in the study area was misclassified.

Lastly, the number of lanes can be missed because of GPS signal loss in tunnels.

Such misclassifications require further investigation to improve the accuracy classification of road segments by number of lanes. In summary, our method performs much better than other methods for number of lanes identification from low-precision GPS trajectories with low-sampling frequency.

## 4.4.2. Quantitative Evaluation for Turn Rules Detection

To evaluate the performance of the proposed method for turn rules of each lane, we compared turn information of two intersections calculated by MLIT with that of manually marked roads. In Figure 14, intersections were randomly selected from a database and the trajectories that traversed those intersections were used to detect turn rules of each lane. The results show that the overall accuracy for turn information classification was 81.3% when comparing our detection results with the actual turn rules, assuming the rate of non-standard driving was 5%. The accuracy of turn rules of each lane identification is lower than the number of lane extraction because it depends not just on the accuracy of lane number identification, but driver behavior and data precision as well.

**Figure 14.** The overlay of image and trajectories. (**a**) shows the tracking results of one intersection; (**b**) indicates the tracking results of another intersection.

## 5. CONCLUSIONS

In this paper, we proposed an automated method (MLIT) to extract lane information, such as numbers of lane and lane turns on road segments from low-precision GPS trajectory data. On one hand, the proposed method (MLIT) eliminates outliers from GPS trajectory data using adaptive density optimization, method improving the robustness of the lane information detection. On the other hand, MILT detects the exact numbers of lanes in TSSes by combining prior knowledge with trace features of road planes and road profiles, resulting in robust extraction of numbers of lanes. However, MLIT still has room for improvement, and the future work will continue to focus on trajectory optimization and extraction of lane information in complex road environments such as tunnels, or overpasses.

## ACKNOWLEDGMENTS

The research presented here was funded by the National Natural Science Foundation of China (No. 41571430, 41271442, 40801155), and the open research fund of the Academy of Satellite application (2014_CXJJ-DSJ_02). We acknowledge Shenzhen Science Technology Bureau for institutional support.

## AUTHOR CONTRIBUTIONS

Luliang Tang and Xue Yang conceived and designed the algorithms of MLIT presented in this paper. Xue Yang performed the experiments and wrote the paper. Zihan Kan and Qingquan Li contributed analysis tools.

## CONFLICTS OF INTEREST

The authors declare no conflict of interest.

## REFERENCES

1. Gonzalez, J.P.; Ozguner, U. Lane detection using histogram-based segmentation and decision trees. In Proceedings of 2000 IEEE

Intelligent Transportation Systems, Dearborn, MI, USA, 1–3 October 2000.
2. Wang, Y.; Teoh, E.K.; Shen, D. Lane detection using B-snake. In Proceedings of 1999 International Conference on Information Intelligence and Systems, Bethesda, MD, USA, 3 October 1999.
3. Hillel, A.B.; Lerner, R.; Levi, D.; Raz, G. Recent progress in road and lane detection: A survey. *Mach. Vis. Appl.* **2014**, *25*, 727–745.
4. Kammel, S.; Pitzer, B. Lidar-based lane marker detection and mapping. In Proceedings of Intelligent Vehicles Symposium, Eindhoven, the Netherlands, 4–6 June 2008.
5. Thuy, M.; León, F. Lane detection and tracking based on Lidar data. *Metrol. Meas. Syst.* **2010**, *17*, 311–321.
6. Yang, B.S.; Dong, Z.; Zhao, G.; Dai, W.X. Hierarchical extraction of urban objects from mobile laser scanning data. *ISPRS J. Phothogr. Remote Sens.* **2015**, *99*, 45–57.
7. Chen, Y.H.; Krumm, J. Probabilistic modeling of traffic lanes from GPS traces. In Proceedings of the 18th SIGSPATIAL International Conference on Advances in Geographic Information Systems, San Jose, CA, USA, 2–5 November 2010.
8. Yeh, A.G.O.; Zhong, T.; Yue, Y. Hierarchical polygonization for generating and updating lane-based road network information for navigation from road markings. *Int. J. Geogr. Inf. Sci.* **2015**, *29*, 1509–1533.
9. Liu, X.T.; Ban, Y.F. Uncovering spatio-temporal cluster patterns using massive floating car data. *ISPRS Int. J. Geo-Inf.* **2013**, *2*, 371–384.
10. Sainio, J.; Westerholm, J.; Oksanen, J. Generating heat maps of popular routes online from massive mobile sports tracking application data in milliseconds while respecting privacy. *ISPRS Int. J. Geo-Inf.* **2015**, *4*, 1813–1826.
11. Zheng, Y.; Zhang, L.; Xie, X.; Ma, W.Y. Mining interesting locations and travel sequences from GPS trajectories. In Proceedings of 18th International World Wide Web Conference, Madrid, Spain, 20–24 April 2009.
12. Giannotti, F.; Nanni, M.; Pinelli, F.; Pedreschi, D. Trajectory pattern mining. In Proceedings of 13th ACM SIGKDD International

Conference on Knowledge Discovery and Data Mining, San Jose, CA, USA, 12–15 August 2007.
13. Yin, P.; Ye, M.; Lee, W.C.; Li, Z. Mining GPS data for trajectory recommendation. In *Advances in Knowledge Discovery and Data Mining*; Springer: Cham, Switzerland, 2014; pp. 50–61.
14. Tang, L.L; Chang, X.M.; Li, Q.Q. Public travel route optimization based on ant colony optimization algorithm and taxi GPS data. *China J. Highw. Transp.* **2011**, *24*, 89–95.
15. Wang, J.; Rui, X.; Song, X.; Tan, X. A novel approach for generating routable road maps from vehicle GPS trajectories.*Int. J. Geogr. Inf. Sci.* **2014**, *29*, 69–91.
16. Tang, L.L.; Huang, F.Z.H.; Zhang, X.Y.; Li, Q.Q. Road Network change detection based on floating car data. *J. Netw.***2012**, *7*, 1063–1070.
17. Zhou, B.D.; Li, Q.Q.; Mao, Q.Z.H.; Tu, W.; Zhang, X.; Chen, L. ALIMC: Activity landmark-based indoor mapping via crowdsourcing. *IEEE Trans. Intell. Transp. Syst.* **2015**, *16*, 2774–2785.
18. De Fabritiis, C.; Ragona, R.; Valenti, G. Traffic estimation and prediction based on real time floating car data. In Proceedings of 11th International IEEE Conference on Intelligent Transportation Systems (ITSC), Beijing, China, 12–15 October 2008.
19. Sun, D.; Zhang, C.; Zhang, L.; Chen, F.; Peng, Z.R. Urban travel behavior analyses and route prediction based on floating car data. *Trans. Lett. Int. J. Trans. Res.* **2014**, *6*, 118–125.
20. Lee, W.C.; Krumm, J. Trajectory preprocessing. In *Computing with Spatial Trajectories*; Zheng, Y., Zhou, X., Eds.; Springer: New York, NY, USA, 2011; pp. 3–33.
21. Brakatsoulas, S.; Pfoser, D.; Salas, R.; Wenk, C. On map-matching vehicle tracking data. In Proceedings of 31st International Conference on Very Large Data Bases, Trondheim, Norway, 30 August–2 September 2005.
22. Haklay, M.; Weber, P. OpenStreetMap: User-generated street maps. *IEEE Perv. Comput.* **2008**, *7*, 12–18.
23. Yanagisawa, Y.; Akahani, J.; Satoh, T. Shape-based similarity query for trajectory of mobile objects. In Proceedings of 4th International

Conference on Mobile Data Management, Melbourne, Australia, 21–24 January 2003.

24. Bruntrup, R.; Edelkamp, S.; Jabbar, S. Incremental map generation with GPS traces. In Proceedings of the 2005 IEEE Intelligent Transportation Systems, Vienna, Austria, 13–15 September 2005.

25. Li, J.; Qin, Q.; Xie, C.; Zhao, Y.; Li, J.; Qin, Q. Integrated use of spatial and semantic relationships for extracting road networks from floating car data. *Int. J. Appl. Earth Obs. Geoinf.* **2012**, *19*, 238–247.

26. Liu, C.H.Y.; Xiong, L.; Hu, X.Y.; Shan, J. A progressive buffering method for road map update using OpenStreetMap data. *ISPRS Int. J. Geo-Inf.* **2015**, *4*, 1246–1264.]

27. Li, Q.Q.; Tang, L.L.; Zuo, X.Q.; Li, H.W. Transect-based three dimensional road modeling and visualization. *Geo-Spat. Inf. Sci.* **2004**, *7*, 14–17.

28. Pollak, K.; Peled, A.; Hakkert, S. Geo-based statistical models for vulnerability prediction of highway network segments. *ISPRS Int. J. Geo-Inf.* **2014**, *3*, 619–637.

29. Wagstaff, K.; Cardie, C.; Rogers, S.; Schroedl, S. Constrained k-means clustering with background knowledge. In Proceedings of 18th International Conference on Machine Learning (ICML), Williamstown, MA, USA, 28 June–1 July 2001.

30. Edelkamp, S.; Schrödl, S. Route planning and map inference with global positioning trajectories. *Comput. Sci. Perspect.* **2003**, *2598*, 128–151.

31. Uduwaragoda, A.; Perera, A.S.; Dias, S.A.D. Generating lane level road data from vehicle trajectories using kernel density estimation. In Proceedings of the 16th International IEEE Annual Conference on Intelligent Transportation Systems (ITSC), Hague, the Netherlands, 6–9 October 2013.

32. Han, J.; Kamber, M. Mining stream, time-series, and sequence data. In *Data mining: Concepts and techniques*; Asma, S., Ed.; Elsevier: USA, 2011; pp. 467–531.

33. Shekhar, S.; Evans, M.R.; Kang, J.M.; Pradeep, M. Identifying patterns in spatial information: A survey of methods. *WIREs Data Min. Knowl. Discov.* **2011**, *1*, 193–214.

34. Liu, Q.; Tang, J.; Deng, M.; Shi, Y. An iterative detection and removal method for detecting spatial clusters of different densities. *Trans. in GIS* **2015**, *19*, 82–106.
35. Lee, J.G.; Han, J. Trajectory clustering: A partition-and-group framework. In Proceedings of the 2007 ACM SIGMOD International Conference on Management of Data, Beijing, China, 11–14 June 2007.

# CHAPTER 6

## SEMI-SUPERVISED BAYESIAN CLASSIFICATION OF MATERIALS WITH IMPACT-ECHO SIGNALS

Jorge Igual, Addisson Salazar, Gonzalo Safont and Luis Vergara

Departamento de Comunicaciones, Universitat Politecnica de Valencia, Camino de Vera s/n, 46022 Valencia, Spain

## ABSTRACT

The detection and identification of internal defects in a material require the use of some technology that translates the hidden interior damages into observable signals with different signature-defect correspondences. We apply impact-echo techniques for this purpose. The materials are classified according to their defective status (homogeneous, one defect or multiple defects) and kind of defect (hole or crack, passing through or not). Every specimen is impacted by a hammer, and the spectrum of the propagated wave is recorded. This spectrum is the input data to a Bayesian classifier that is based on the modeling of the conditional probabilities with a mixture of Gaussians. The parameters of the Gaussian mixtures and the class probabilities are estimated using an extended expectation-maximization algorithm. The advantage of our proposal is that it is flexible, since it obtains good results for a wide range of models even under little supervision; e.g., it obtains a harmonic average of precision and recall value of 92.38% given only a 10% supervision ratio. We test the method with real specimens made of aluminum alloy. The results show that the algorithm works very well. This technique could be applied in many industrial problems, such as the optimization of the marble cutting process.

## KEYWORDS

impact echo; accelerometers; mixture of Gaussians; semi-supervised Bayes classification

## 1. INTRODUCTION

The field of non-destructive testing (NDT) of materials is a wide area, including any technique that extracts information about the condition of a material specimen without altering its physical and/or chemical properties (see, e.g., [1] for a survey of different NDT methods).

Two main elements appear in NDT: sensors and data processing. While sensors are very application dependent and impose practical limits about monitoring resolution, data processing considers general techniques, which may find application in a variety of significantly different NDT problems. From another perspective, sensors are limited by the current sensor technology; meanwhile, data processing is only limited by the required computational resources. On the other hand, non-destructive methods often lead to automatic implementations, thus allowing "on-line" monitoring of large amounts of specimens.

The sensory system poses the essential resolution limits that can be reached to measure the material state. However, large improvements can be achieved in the overall system performance by improving the data processing methods. Although based on general techniques, these methods must take into account the specific context where the NDT is to be applied. Hopefully, once success is demonstrated in that specific context, the method could be extended to other significantly different NDT problems.

In this paper, we present a new classification method, which is specifically oriented to scenarios where some degree of supervision is allowed. The general goal is to classify the specimen under analysis in one of a predefined number of classes. The classifier is trained on the basis of a set of feature vectors previously computed using the same NDT method and sensors. A subset of the whole set is labeled, while the rest is considered of an unknown class. This is termed semi-supervised learning [2] and makes sense in those NDT problems where some selected specimens can be "*a posteriori*" analyzed in a destructive manner, so that the "true" class of the specimen could be known to train the classifier of future specimens incoming to the system.

We apply the proposed data processing in the context of one particular NDT method: impact-echo (IE) [3]. In this technique, a material is impacted with a hammer, which produces an acoustic response that is collected by the sensory system located on the surface of the material. Usually, a set of sensors are distributed across the different sides of the specimen to extract exhaustive information about its inner state. The underlying physics is that of acoustic wave propagation in solids, where different types of waves propagates into the solid and are recorded by properly-selected sensors when they arrive at the surface. The waves are P-wave (normal stress), S-wave (shear stress) and R-wave (surface or Rayleigh). Then, signal processing is performed on the collected signals to extract features that, grouped in a vector form, are the inputs to the automatic classifier subsystem. The feature vector is preprocessed using principal component analysis (PCA) [4]. PCA allows reducing the dimension of the input feature vector while retaining most of the variation in the original data. It is used in many classification problems, including the IE field [5].

IE is a low-resolution technique, which has been extensively applied to monitor the general state of specimens. It has attractive advantages, like low cost, rapid global analysis, deeppenetration and "on-line" processing capability. It is not appropriate for exact localization or characterization of inner defects, where other higher resolution NDT methods, like ultrasonics, are more adequate. However, we will show in the experimental part of this paper that some specific information about the inner state can be obtained by properly defining the targeted classes of the classifier.

Signal processing in IE can be roughly organized into four classes depending on the assumptions considered: time domain, frequency domain, time-frequency domain and machine learning. The first IE works were in the time and frequency domain. Time domain analysis is based on the estimation of successive P-wave arrivals (multiple reflections between the parallel surfaces of a plate) that allows the period and dominant frequency of the waveform to be estimated. In practice, the conditions of an ideal plate are difficult to reach, and thus, a quick interpretation of the results in the time domain is also difficult [6]. In frequency domain analysis, the fast Fourier transform (FFT) is used to obtain the spectrum of the impact-echo signal. The value of the maximum peak frequency in the amplitude spectrum is used to determine the thickness of the plate (see, e.g., [7]).

Spectral analysis of IE signals was improved using time-frequency techniques considering their non-stationarity, *i.e.*, the transient nature of the IE signals. The principal aim was to overcome the problem of noisy signals where the reflections are not clearly distinguished in the spectrum, which shows multiple peaks due to artificial energy added by relatively strong R-waves. This is particularly pronounced in cases of limited dimension specimens. Several time-frequency techniques, such as short-time Fourier transform (STFT) and Hilbert–Huang, have been applied to improve the accuracy in thickness estimation [8,9]. Recently, systematic errors in thickness estimation from IE testing due to near-field effects on the P-wave and R-wave were investigated [10].

The ultimate advances in IE signal processing research came from the field of machine learning and statistical pattern recognition. These methods extract some features from signals of specimens of known classes and use them to train a pattern recognition algorithm that can be used to classify other specimens of an unknown class. Several NDT applications can be suited to this framework, for instance the classification of a material depending on the kind and number of defects, which is the problem addressed in this paper. The degrees of freedom afforded by this framework facilitate multichannel analysis, simultaneous use of features from different domains and the combination of different NDT methods.

Some examples of the combination of IE with other NDT methods to improve the results of defect detection problems are the following: combination with the impulse-response method for identifying delaminations in concrete floor toppings [11] and combination with ultrasonic pulse echo and ground penetrating radar data (GPR) for detecting built-in honeycombing in scale concrete specimens [12].

The machine learning methods most commonly applied in IE signal analysis are based in artificial neural networks (ANN). The problems studied with these methods include: prediction of the concrete compressive strength and thickness of concrete structures [13]; prediction of the internal grouting quality of prestressed ducts [14]; and identification of the pull-off adhesion of the concrete layers in floors on the basis of parameters evaluated on the structural layer surface [15]. Recently, a linear subspace representation of the original features, called the Grassmann manifold, was applied in IE. It was demonstrated that subspace representation could characterize relevant time-frequency distribution patterns and form significant clusters that are separable using a distance [5].

The machine learning method proposed herein has the following advantages compared with the other methods mentioned above: (I) enabling semi-supervised learning (capable of incorporating different proportions of unlabeled and labeled data); this facilitates a quick implementation of the method with a very small sample of specimens, faster than other supervised methods, such as ANN (the difficulties and cost of obtaining labeled data have been extensively studied; see, e.g., [2]); (II) the level of operation of the classification system can be adjusted depending on the percentage of false alarms allowed; (III) the proposed multichannel setup allows mass spectra to be captured from the IE testing experiments that register the differences between defective and homogeneous kinds of materials; (IV) the advantages of a generative model; we obtain posterior probabilities for every class, so this probability can be used in many different ways, not only for basic classification purposes, such as the maximum *a posteriori* (MAP) estimate.

IE has been applied in different types of materials, such as marble, concrete or steel [11,16–18]. In those cases, large blocks are inspected to ascertain the general quality before cutting the material into slabs. This prevents the possibility of accidents during the cutting process, which can deteriorate the machinery and be dangerous to the human operators. It also helps in the setting of the block quality, *i.e.*, in the final price of the material. Training a block classifier is possible in this type of application by selecting a set of blocks where training feature vectors are obtained using the IE method. Some of these blocks are carefully inspected after cutting to judge the true inner state. In practice, this can be done only in a small number of blocks, so we have a semi-supervised scenario.

The method proposed in this paper assumes knowledge of the number of classes. The multivariate probability density of the feature vectors corresponding to a particular class is considered to be a mixture of Gaussians (MoG). An MoG, also referred to as a Gaussian mixture model (GMM), assumes that all of the feature vectors for a given class are generated from a weighted sum of a finite number of Gaussian distributions with unknown parameters (mean and covariance). Hence, every feature vector is generated by one of the mixture components of a given class.

A given feature vector can be originated, in principle, by any component of any of the classes. However, labeled features are known to be generated by one of the components corresponding to the labeled class, although the specific component inside the class is unknown. This knowledge can be incorporated into the estimation of the whole model parameters, thus improving the performance of a Bayes classifier.

In Section 2, the new semi-supervised method is presented. Then, in Section 3, the dataset and the IE experimental setup are described. In Section 4, we present exhaustive results considering different target classes and levels of supervision, followed by the conclusions.

## 2. METHOD

There are two basic paradigms in order to define a classifier: a discriminative approach, where the goal is to directly assign the observations to the correct class, obtaining a rule that tries to minimize the errors, and a generative model approach, where we try to learn how the observations are generated, and after that, we assign the observation to the model with the highest probability. We will follow this second approach using the Bayesian classifier.

The Bayesian classifier calculates the posterior distribution $p(k/x)$ for every class $k$, $k = 1 \ldots K$ and labels the observation $\mathbf{x} = [x_1, \ldots, x_d]^T$ with the class that has the largest probability. Applying Bayes' rule, we obtain:

$$p(k/\mathbf{x}) = \frac{p(\mathbf{x}/k)p(k)}{\sum_{k=1}^{K} p(\mathbf{x}/k)p(k)} \tag{1}$$

where $p(x/k)$ is the conditional probability density of an observation vector x for class $k$, $k = 1, 2, \ldots, K$ and $p(k)$ is the corresponding prior distribution for every class with $\sum_{k=1}^{K} p(k) = 1$.

The conditional distributions are modeled by an MoG for every class. An MoG is a weighted sum of Gaussians with mean $\mu_i$ and covariance matrix $\Sigma_i$:

$$p(\mathbf{x}/k) = \sum_{i_k=1}^{I_k} \alpha_{i_k} N_{i_k}(\mathbf{x}; \mu_{i_k}, \Sigma_{i_k})$$
$$N_{i_k}(\mathbf{x}; \mu_{i_k}, \Sigma_{i_k}) = (2\pi)^{-d/2} |\Sigma_{i_k}|^{-1/2} e^{-\frac{1}{2}(\mathbf{x}-\mu_{i_k})^T \Sigma_{i_k}^{-1}(\mathbf{x}-\mu_{i_k})}$$

$$\tag{2}$$

where $I_k$ is the number of Gaussians used in the MoG that models the conditional distribution of class $k$.

For every class, each Gaussian contributes to the mixture model in the proportion or mixing coefficient $\alpha_{ik}$, with $\alpha_{ik} \geq 0$ and $\sum{ik=1}Ik\alpha ik=1$. These weights can also be interpreted as priors, indicating the prior probability of the data coming from the corresponding Gaussian of the mixture.

When an observation x is available, we can apply Bayes' theorem to calculate the posterior probability in Equation (1), *i.e.*, the probability that the observation comes from each class, and classify accordingly.

However, we need to estimate the previous model parameters. They include the prior probabilities of every class and the parameters of the different mixture models. In order to do this, we define the log-likelihood function $L(\mathbf{X})$ of $N$ observations $\mathbf{X} = [x_1, ..., x_N]$:

$$L(\mathbf{X}; \Psi) = \log p(\mathbf{X}; \Psi) = \sum_{n=1}^{N} \log p(\mathbf{x}_n; \Psi) = \sum_{n=1}^{N} \log \sum_{k=1}^{K} p(\mathbf{x}_n/k)p(k)$$

(3)

with the set of parameters to be estimated $\Psi = \{\Psi_1, ..., \Psi_k\}$, where $\Psi_k = \{p(k), \alpha_{ik}, \mu_{ik}, \Sigma_{ik}\}$.

Using Equation (2) in Equation (3), we obtain:

$$L(\mathbf{X}; \Psi) = \sum_{n=1}^{N} \log \sum_{k=1}^{K} p(k) \sum_{i_k=1}^{I_k} \alpha_{i_k} N_{i_k}(\mathbf{x}_n; \mu_{i_k}, \Sigma_{i_k})$$

(4)

The maximum likelihood estimator calculates the parameters that maximize Equation (4), *i.e.*, $\Psi MLE = \arg\max_\Psi L(X;\Psi)$. Since this equation involves the log of a sum, it is not easy to find the maximum, and an expectation-maximization (EM) approach [19] is better suited.

As usual, we assume that the observations $\mathbf{X}$ are part of a complete dataset $(\mathbf{X}, \mathbf{Z})$, $\mathbf{Z} = [\mathbf{z}_1, ..., \mathbf{z}_N]$, where $\mathbf{z}_n$ is a random vector of dimension $I = \sum_{k=1}^{K} I_k$. This vector is equal to zero, but one element, which is equal to one, the class and Gaussian component that is responsible for the observation $x_n$. i.e.,:

$$z_n = [\underbrace{0,\ldots 0}_{I_1},\ldots,\underbrace{0,\ldots,1,\ldots 0}_{I_k},\ldots,\underbrace{0,\ldots 0}_{I_K}] \qquad (5)$$

with the one corresponding to the *m*-th element, *i.e.*, the *k*-th class and $i_k$-th component in the mixture model of the *k*-th class conditional distribution.

The corresponding complete data log-likelihood reads:

$$L_c(\mathbf{X}, \mathbf{Z}; \Psi) = \log p(\mathbf{X}, \mathbf{Z}; \Psi) = \sum_{n=1}^{N} \log p(\mathbf{x}_n, \mathbf{z}_n; \Psi) \qquad (6)$$

Using Equation (5) $L_c(\mathbf{X}, \mathbf{Z}; \Psi)$ can be expressed as:

$$L_c(\mathbf{X}, \mathbf{Z}; \Psi) = \log \prod_{n=1}^{N} \prod_{m=1}^{I} (p(\mathbf{x}_n/z_{nm}=1)p(z_{nm}=1))^{z_{nm}} =$$
$$\sum_{n=1}^{N} \sum_{m=1}^{I} z_{nm}(\log p(\mathbf{x}_n/z_{nm}=1) + \log p(z_{nm}=1)) \qquad (7)$$

Following the EM procedure, we first take the expectation of the complete log-likelihood. This expectation is obtained assuming that the model parameters are fixed $\Psi = \Psi^j$, *i.e.*, the class probabilities, proportions, means and covariance matrices are fixed. If the expectation is taken with respect to the posterior distribution of the unobserved data, it is guaranteed that maximizing the complete log-likelihood, we are also maximizing the incomplete log-likelihood function, which is the real problem. This is the expectation step in the EM algorithm and can be summarized as:

$$Q(\Psi; \Psi^j) = E[\log p(\mathbf{X}, \mathbf{Z}; \Psi)/\mathbf{X}, \Psi^j] \qquad (8)$$

$$Q(\Psi; \Psi^j) = \sum_{n=1}^{N}\sum_{m=1}^{I} E[z_{nm}](\log(p(\mathbf{x}_n/z_{nm}=1) + \log \delta_m) \qquad (9)$$

where $\delta_m = p(z_{nm}=1)$, with the constraint $\sum_{m=1}^{I}\delta_m = 1$.

The expectation step calculates the expected value of the responsibilities using the present values of the parameters:

$$E[z_{nm}/\mathbf{x}_n, \Psi^j] = 1 \cdot p(z_{nm}=1/\mathbf{x}_n, \Psi^j) + 0 \cdot p(z_{nm}=0/\mathbf{x}_n, \Psi^j) =$$
$$= p(z_{nm}=1/\mathbf{x}_n, \Psi^j) \qquad (10)$$

The posterior probabilities $p(z_{nm}=1/\mathbf{x}_n, \Psi^j)$, i.e., the responsibility that the element $m$-th takes for generating the $n$-th observation given the current model parameters, are obtained using Bayes' theorem:

$$p(z_{nm}=1/\mathbf{x}_n, \Psi^j) = \frac{p(\mathbf{x}_n/z_{nm}=1)p(z_{nm}=1)}{\sum_{m=1}^{I} p(\mathbf{x}_n/z_{nm}=1)p(z_{nm}=1)} = \frac{p(\mathbf{x}_n/z_{nm}=1)\delta_m}{\sum_{m=1}^{I} p(\mathbf{x}_n/z_{nm}=1)\delta_m} \qquad (11)$$

Once the expectation is calculated, Equation (7) is no longer a random variable; we have just a log-likelihood function $Q(\Psi; \Psi^j)$ that can be maximized as usual with respect to the model parameters, obtaining a new estimate of them $\Psi^{j+1}$; this is the maximization step of the algorithm:

$$Q(\Psi; \Psi^j) = \sum_{n=1}^{N}\sum_{m=1}^{I} p(z_{nm}=1/\mathbf{x}_n, \Psi^j)(\log(p(\mathbf{x}_n/z_{nm}=1) + \log \delta_m) \qquad (12)$$

This two-step procedure is repeated iteratively until convergence, and it is guaranteed that in every iteration, the likelihood function is increased and, therefore, converges to a local maximum.

To obtain the new parameters, we have to maximize $Q(\Psi; \Psi')$. In order to maximize $Q(\Psi; \Psi')$, we take derivatives with respect to every parameter of the model and set them equal to zero. Note that the function $Q(\Psi; \Psi')$ can be decoupled in a sum of different terms, where each of them includes only one kind of parameter, i.e., $\delta_m^{j+1}, \mu_m^{j+1}, \Sigma_m^{j+1}$. Remember that index $m$ is related to the corresponding $i_k$ in Equation (2) depending on the number of classes and Gaussians per class model.

Once the objective function $Q(\Psi; \Psi')$ is decomposed, we can obtain the new mean, variance and component weight in the same way as in the classic EM algorithm:

$$\mu_m^{j+1} = \frac{\sum_{n=1}^{N} p(z_{nm} = 1/\mathbf{x}_n, \Psi^j) \mathbf{x}_n}{\sum_{n=1}^{N} p(z_{nm} = 1/\mathbf{x}_n, \Psi^j)} \tag{13}$$

$$\Sigma_m^{j+1} = \frac{\sum_{n=1}^{N} p(z_{nm} = 1/\mathbf{x}_n, \Psi^j)(\mathbf{x}_n - \mu_m^j)(\mathbf{x}_n - \mu_m^j)^T}{\sum_{n=1}^{N} p(z_{nm} = 1/\mathbf{x}_n, \Psi^j)} \tag{14}$$

$$\delta_m^{j+1} = \frac{1}{N} \sum_{n=1}^{N} p(z_{nm} = 1/\mathbf{x}_n, \Psi^j) \tag{15}$$

We estimate the new probability for every class $p(k^{j+1})$ integrating out the corresponding δj+1m elements:

$$p(k^{j+1}) = \sum_{m=m_0+1}^{M} \delta_m^{j+1} \tag{16}$$

where the indexes of the summation are:

$$m_0 = \sum_{j=1}^{k-1} I_j$$
$$M = \sum_{j=1}^{k} I_j \tag{17}$$

The new weights for every Gaussian component are obtained normalizing the full responsibility by the corresponding class probability:

$$\alpha_m^{j+1} = \frac{\delta_m^{j+1}}{p(k^{j+1})} \tag{18}$$

In the case that all classes have the same number of Gaussians, the notation simplifies to:

$$p(k^{j+1}) = \sum_{m=(k-1)R+1}^{kR} \delta_m^{j+1}, \quad k = 1, \ldots, K \tag{19}$$

$$\alpha_m^{j+1} = \frac{\delta_m^{j+1}}{\sum_{n=\lfloor m/R \rfloor R+1}^{\lceil m/R \rceil R} \delta_n^{j+1}}, \quad m = 1, \ldots, KR \tag{20}$$

where $\lfloor . \rfloor$, $\lceil . \rceil$ are the floor and ceiling operators, respectively.

Supervision is introduced implicitly in Equation (11). The posterior probability $p(z_{nm} = 1/x_n, \Psi^j)$ for samples with a known class $k$ is easily computed, such as:

$$p(z_{nm} = 1/\mathbf{x}_n, \Psi^j) = \frac{p(\mathbf{x}_n/z_{nm} = 1)\alpha_m}{\sum_{m=1}^{I} p(\mathbf{x}_n/z_{nm} = 1)\alpha_m} \tag{21}$$

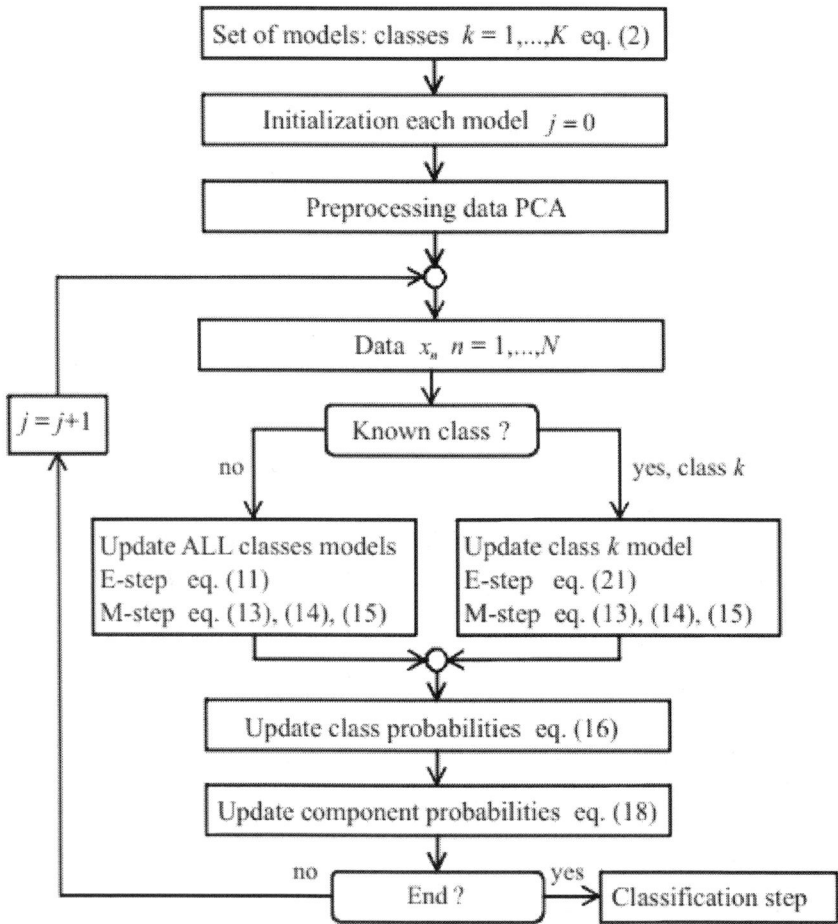

**Figure 1.** Flowchart of the algorithm.

for the interval of indexes $m$ corresponding to the components of the mixture model of the known class $k$ with priors $\alpha_m$ and $p(z_{nm} = 1/x_n, \Psi^j) = 0$ for the rest of indexes $m$.

Since this posterior probability is used in the maximization step, it means that samples $x_n$ that belong to a known class $k$ are used to update the parameters of that class only: mean, covariance and prior probabilities of the MoG for that class; see Equations (13)–(15). In other words, the rest of the classes do not take into account those samples in the updating step of their parameters, since in their sums, those terms are zero, $p(z_{nm} = 1/x_n, \Psi^j) = 0$, for indexes $m$ out of the interval corresponding to the known class $k$.

The algorithm is summarized in Figure 1. First, you set the models (one MoG per class) and the parameters (how many Gaussians per class). Second, the mean and covariance matrices of each class are initialized. For this purpose, only the samples from a known class (supervised samples) are used in the initialization of the corresponding class parameters. Third, the data are preprocessed using PCA in order to reduce the dimensions of the feature vector. Fourth, the algorithm is run until convergence. The algorithm stops when the new and old parameters change less than a threshold value.

## 3. DATA

We apply the explained algorithm to real data obtained in the lab using materials made of aluminum alloy series 2000 of dimensions $7 \times 5 \times 22$ cm (width, height and length, respectively). These dimensions were appropriate for lab experiments and may be considered reasonable scale replicas of real specimens used in different problems were the impact echo method has been applied (see, for example, [20–22] and the references therein). Moreover, these dimensions are appropriate for a dense excitation of resonant modes [23], thus leading to rich spectrum content. We show an example of a piece under study in Figure2: arrows point to the hammer and the accelerometers, while the red and white cables connect the accelerometers to the acquisition equipment.

Up to three defects per piece were drilled in different locations of each piece. The defects passed through the pieces and consisted of holes in the shape of cylinders of 10 mm and cracks in the shape of parallelepipeds of 5 × 20 mm cross-sections. Some of the defects cross all of the pieces, e.g., a hole that passes totally through the other face of the material, and some others do not, stopping at some point in the interior of the piece. The material was excited by an impact, and its response was measured by the accelerometers (sensors).

As an example, in Figure 3, we show the setup for a piece with a hole and a crack defect. We use seven sensors located on different surfaces of the parallelepiped in order to capture the information coming from different directions and distances from the impact and defects. In the example provided in the figure, the hammer impacts on the front face. There is a hole in the Y axis far from the impact surface and one crack in the XZ plane near the impact plane.

142  Dynamic Programming and Bayesian Inference, Concepts and Applications

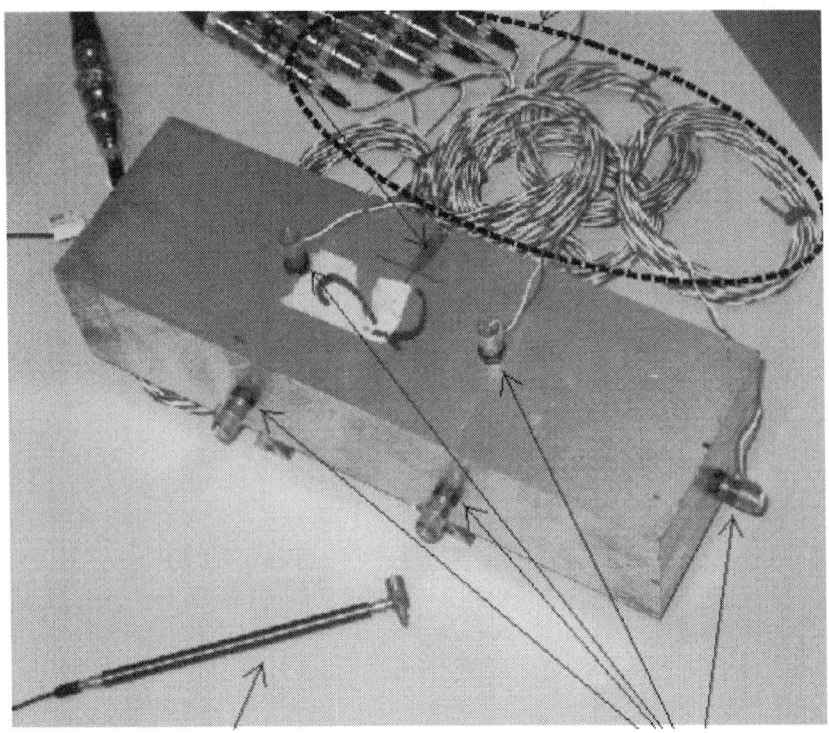

**Figure 2.** Example of the piece under study.

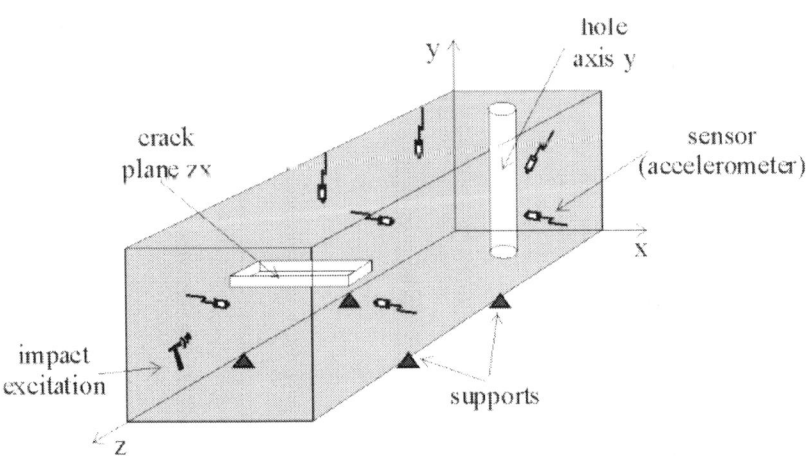

**Figure 3.** Setup experiment for a piece with a hole and a crack.

With respect to the equipment, we used an impact hammer 084A14 PCB, eight accelerometers (a1–a8 in the figure) 353B17 PCB, an ICP signal conditioner F482A18 and a data acquisition module 6067E. The acquisition parameters were: sampling frequency = 100,000 kHz and observation time = 50 ms. We use as input data the spectrum of the recorded signals coming from the sensors, normalized using the maximum of the impact signal amplitude. The total number of experiments included 1881 executions of the IE test from 76 specimens.

The 76 pieces can be grouped into different classes attending to several criteria: the status of the piece, the kind of defect, the orientation of the defect and the length of the defect. Depending on the criteria we use, we can state different classification problems with an increasing number of classes. The first problem has four classes, since the pieces are divided into four groups: non-defective (also called homogeneous), one hole defect, one crack defect and multiple defects.

If we split the one defect pieces into subclasses attending to the orientation of the defect, we have a second problem with eight classes: homogeneous, one hole in the X axis, one hole in the Y axis, one hole in the Z axis, one crack in the XY plane, one crack in the YZ plane, one crack in the XZ plane and multiple defects.

The most challenging case is when we also use the length of the defect: if it goes through the piece or just up to some point in between, we call these subclasses passing through and not passing through, respectively. Thus, in the most complex case, we have fourteen classes.

Due to technical reasons and for simplicity in the making of the specimens in the lab, we have no Z direction hole pieces, nor obviously passing through and not passing through Z direction hole samples. Therefore, the three classification problems that we address have four, seven and twelve classes, respectively.

All of the information about the specimens and data collection is summarized in Table 1.

One important issue in any classification procedure is to take care of the dimensions of the data during the preprocessing of the data. Since we have to estimate some parameters, it is important to be sure that the estimates are accurate enough to prevent possible overfitting. In our case, this means that we have to reduce the dimensions of the feature vector: the spectrum of the signals captured by the accelerometers. To this end, we preprocess the data applying PCA, as explained in [24]. The number of PCA

**Table 1.** Dataset: number of pieces and experiments per class.

| Type of Defect | Number of Pieces | Number of Experiments | Size of Defect (mm) |
|---|---|---|---|
| Homogeneous | 6 | 200 | - |
| Passing through a hole in the X axis | 4 | 85 | $\emptyset = 10$, X:70 |
| Half-passing through a hole in the X axis | 4 | 89 | $\emptyset = 10$, X:35 |
| Passing through a hole in the Y axis | 4 | 84 | $\emptyset = 10$, X:50 |
| Half-passing through a hole in the Y axis | 4 | 83 | $\emptyset = 10$, X:25 |
| Passing through a crack in the XY plane | 8 | 170 | $X = 20, Y = 50, Z = 5$ |
| Half-passing through a crack in the XY plane | 8 | 160 | $X = 20, Y = 25, Z = 5$ |
| Passing through a crack in the ZY plane | 8 | 187 | $X = 5, Y = 50, Z = 20$ |
| Half-passing through a crack in the ZY plane | 8 | 160 | $X = 5, Y = 25, Z = 20$ |
| Passing through a crack in the ZX plane | 8 | 185 | $X = 70, Y = 5, Z = 20$ |
| Half-passing through a crack in the ZX plane | 8 | 182 | $X = 35, Y = 5, Z = 20$ |
| Multiple defects | 6 | 296 | Combinations of cracks and holes |
| Total | 76 | 1881 | |

components that we keep is given by a threshold on the fraction of variance captured by those components.

Another important factor is the number of samples that are available, *i.e.*, the total amount of data available during the learning process. On the one hand, the pieces made in the lab were submitted to different impacts in order to increase the number of samples and to introduce some randomness and noise in the recording process, since the impact is slightly different in every experiment. On the other hand, since the mechanical process of making pieces with specific defects is difficult, we use resampling techniques to increase the size of the sample when necessary [25]. This consists of generating new realizations, called replicas, by adding to the real recorded value a small amount of white Gaussian noise with a small standard deviation. This helps to improve the learning process, and by using cross-validation methods, we can assure that no overfitting problems arise.

## 4. RESULTS

We applied the classifier to the dataset explained in the previous section. We split the samples into two groups: a training set containing 80% of the data and a testing set with the rest of the samples. In order to cross-validate the results, we ran the algorithm 40 times for every experiment with different training-test data, and we show the calculated mean values.

### 4.1. Measures

To quantify the results, we use a confusion matrix [26]. We define the confusion matrix as a matrix where every row represents the estimated class (the result of our algorithm) and every column the true class (the solution); note that the transpose definition (exchanging rows and columns) could have also been used. With $K$ classes, the confusion matrix is $K \times K$, where the diagonal entries correspond to the correct classifications. The off-diagonal values in every row tell us how the wrong classifications for that estimated class (false positives) are distributed among the true classes and the off-diagonal values in every column how many specimens from a given class are wrongly assigned to each of the other classes (false negatives). This matrix contains all of the information about the performance of the algorithm for our dataset, but it can be tedious to analyze the results and obtain simple conclusions from the confusion matrix. Therefore, we will use also other measures obtained from the confusion matrix to clarify the results.

Since we are considering the same cost for every wrong classification, we are not interested at this point in comparisons between particular classes. Therefore, we can obtain $C_i$, $i = 1 \ldots K$ confusion matrices $2 \times 2$, where for each $C_i$ matrix, the $i$-th class is the positive one and the aggregate of the rest of them is the negative one (errors). These matrices are easily obtained by simply summing up the corresponding values of the whole confusion matrix. The $C_i$ matrices allow us to obtain the precision $p_i$ and recall $r_i$ values for every class $i = 1 \ldots K$:

$$p_i = \frac{TP_i}{TP_i + FP_i} \tag{22}$$

$$r_i = \frac{TP_i}{P_i} \tag{23}$$

where $TP_i$ are the true positives (when the estimated and true classes are the same $C_i$), $FP_i$ are the false positives (when we assign the piece to the positive class $C_i$ erroneously, no matter which is the true class) and $P_i$ is the total number of specimens of the $i$-th class. In other words, precision indicates the ability of the algorithm to distinguish between true and false positives, i.e., how many of the pieces assigned to class $i$ are correct; a large $p_i$ value indicates that most of the pieces that are assigned to class $i$ actually belong to that class. Recall, also called the true positive rate or sensitivity, indicates the ability to detect the positive cases, i.e., how many of the pieces from the $i$-th class are detected; a large $r_i$ value indicates that most of the pieces from that class are identified. We can even reduce these two indices to just one, combining them properly; e.g., the $F$ measure, which is defined for every class as the harmonic average of precision and recall values:

$$F_i = \frac{2}{\frac{1}{p_i} + \frac{1}{r_i}} \tag{24}$$

The values of precision, recall and the F measure are between zero and one, with larger values indicating better performance. In order to make the comparison easier between results, we will use percentage values, *i.e.*, in the interval 0–100. It is important to remark that these values must be used with care, especially in cases where the class distributions change or are skewed and when the cost functions are not 0–1 loss functions (no cost to correct classification and the same cost for all errors). In our experiments, we do not have these problems, since we use the same equiprobable class distributions during the training and testing stages, and as we mentioned, we will consider the same cost for any wrong classification.

## 4.2. General Results

In this subsection, we analyze the general behavior of the algorithm no matter which of the three problems we are solving. We are not interested in how the algorithm performs for any particular class, but we want to extract general conclusions about the performance of the algorithm and how it is influenced by different variables, such as the model parameters or the data size.

The first thing we have to establish is the dimension of the feature space, *i.e.*, the number of principal components that we are going to keep after PCA. We will use the 12-class problem to determine the feature space dimension, since it is the most complicated case. To analyze the influence of the dimension of the feature vector, we run the algorithm for different dimensions. We obtain the confusion matrix and then calculate the F value. In Figure 4, we show the box and whiskers plot of the F value for all of the classes and the overall F mean value (40 runs for each one) when the feature vector is a $3 \times 1$ vector (top), a $7 \times 1$ vector (middle) and a $16 \times 1$ vector (bottom). These values correspond to keeping 25%, 50% and 75% of the total amount of the variance when applying PCA. The overall F mean values are: 83.09, 92.38 and 88.46, respectively.

As we can see in Figure 4, the best results are obtained in the middle case, when the data are projected to a seven-dimensional PCA space. When we increase the dimensions from three to seven, the results improve dramatically: the mean F value goes from 83.09 to 92.38. However, if we increase the number of descriptors to 16, then the results for the test data are worse, as shown in the bottom plot (F reduces to 88.46). This fact is especially clear for Classes 3, 8 and 9, with a large variance and a poor mean classification performance. These results correspond to the case of nine Gaussians per class and a 0.3 supervision ratio. However, we obtain the same conclusion for a wide range of numbers of Gaussians ($I_k$ in Equation

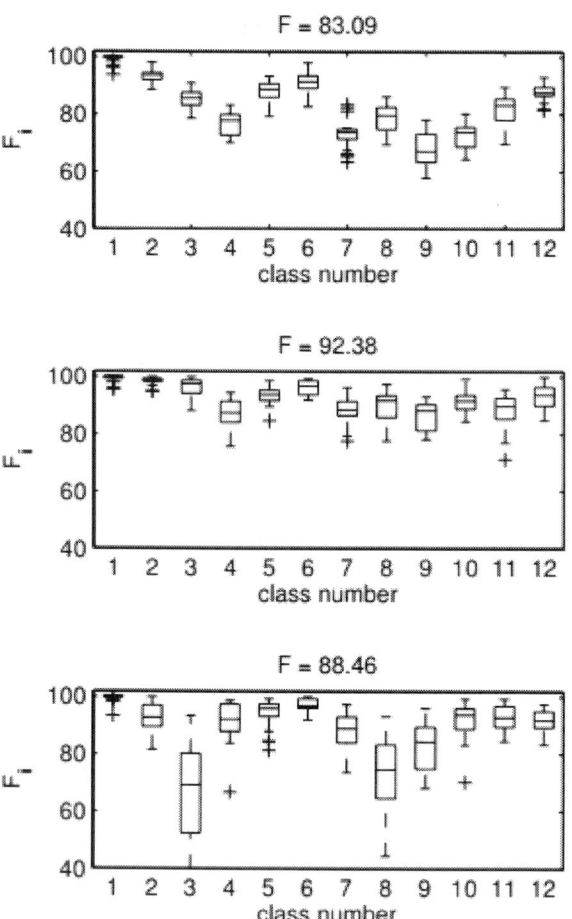

**Figure 4.** Dimension reduction of the feature vector after PCA. Box and whiskers plot of the F measure for each of the 12 classes. (**Top**) 3 × 1 feature vector; (**Middle**) 7 × 1 feature vector; (**Bottom**) 16 × 1 feature vector. Above each plot is the overall mean F measure.

(2)) and supervision values (percentage of known labels used in the training stage). Therefore, we will use the first seven principal components of the PCA transformation of the spectrum as the feature vector.

Once the preprocessing is done, we address the influence of the model complexity on the performance of the classifier, *i.e.*, how to choose the number of Gaussians per class. We seek a number of Gaussians large enough to capture the distribution of the class, but not too high to avoid overfitting. We assume that the complexity of every class is similar, so we will use the same number of Gaussians per class. In such a way, we will not bias the results in favor of any class. With respect to the initialization of the parameters of each Gaussian component for every class (the mean, covariance matrix and proportions of every component), we use the set of supervised samples. The mean of every component corresponds to a random known value of the corresponding class, and the initial covariance matrix is the same for all of the components in the same class: it is obtained as the sample covariance from the known samples, and the components are equiprobable. The initial class priors are calculated obtaining the proportion of every class in the total number of samples with a known class. We used a similar number of samples from every class in all of the problems, so we do not have to worry about the influence of the prior in the evaluation of the results.

We trained and tested the model for an interval of Gaussians, from just three up to 25, in steps of two, $i.e., I_k = 3, 5, \ldots, 25$. In Figure 5, we show the F value obtained for the problem of seven classes during the training (top) and test (bottom) phases. The training results improve as the complexity of the model does, since we can always introduce new Gaussians in order to fit the data better. However, it is clear that the algorithm suffers from overfitting for a number of Gaussians approximately greater than nine, since the test performance is reduced while the training improves; $i.e.$, for a large number of Gaussians, the model is so complex, that it can fit to the training data, but is not able to generalize to new data. To avoid this problem, we restrict our model to a maximum of nine Gaussians per class.

The supervision ratio depends on the available data. It is expected that if the number of samples with a known class increases, the performance of the algorithm should be better. Note that in the extreme case where we know which class any training sample belongs to, our algorithm reduces to the estimation of every class mixture model through the standard EM algorithm. Once the different distributions are obtained, it would be easy to assign the new data to the class with higher probability. On the other hand, for the unsupervised case, every sample will contribute to every mixture model, so it could become an unsolvable problem if modes of different class distributions overlap, because it would be impossible to assign it to one model or another. The algorithm could find the correct Gaussian components in the overall observation distribution, but it could assign the components to the wrong class. If this is the case, the only solution is to introduce some level of supervision in order to guarantee that the components are assigned properly to the corresponding class. In addition, as explained previously, we

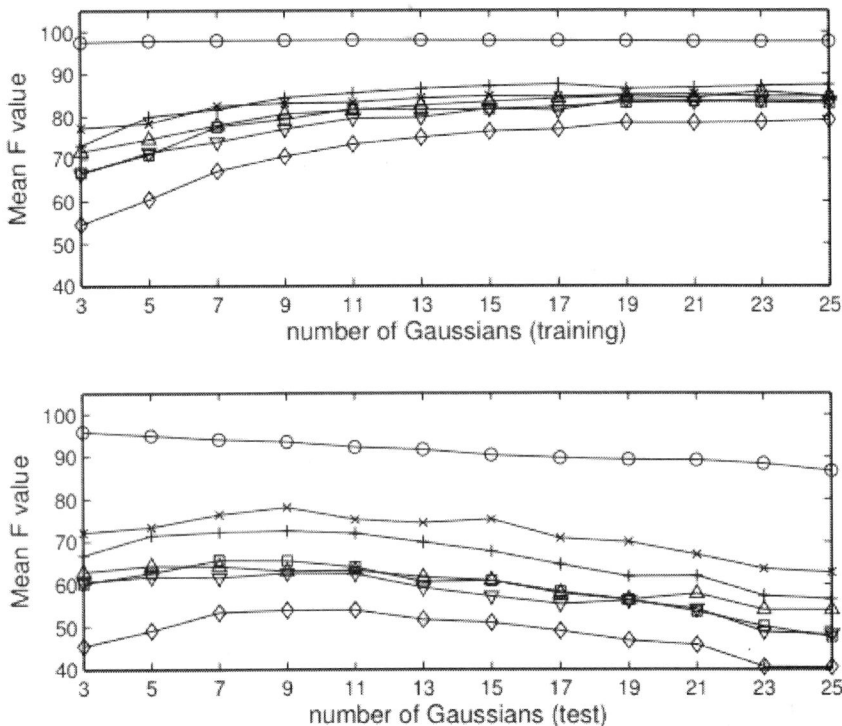

**Figure 5.** Model selection: number of Gaussians per class for the seven classes problem. **(Top)** Mean F value of each class for the training dataset *vs.* the number of Gaussians per class; **(Bottom)** the same for the testing dataset.

cannot afford to have a very large number of samples of every class, since it would be very expensive and time consuming. That is to say, it would be a nice feature of the classifier to learn from a reduced number of samples. Therefore, we need to test the influence of the supervision ratio and sample size in the performance of the algorithm in order to quantify this effect.

To analyze the influence of the data size and supervision, we obtain the results for different sets of samples per class, 50, 100, 200, 300 and 500, and two different supervision rates, 0.1 and 0.5. In Figure 6, we show the results for the four- and 12-class problems. Again, we use the F value in the figures to simplify the analysis.

As we can see, the algorithm needs a minimum amount of data of around 200 samples per class to obtain good results in both problems, *i.e.*, the sample size matters: if it is very low, e.g., 50 or 100, the classifica-

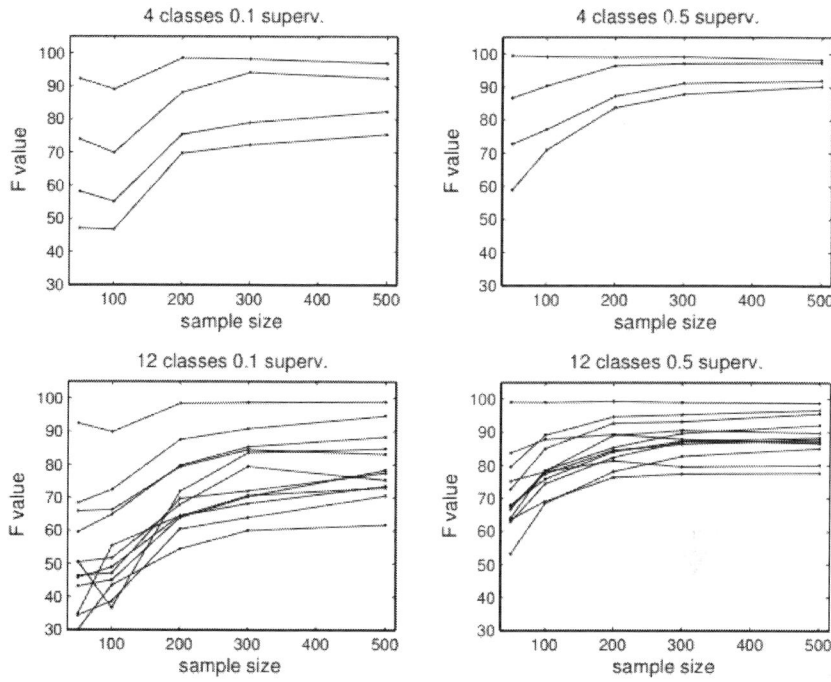

**Figure 6.** Influence of data size and supervision. (**Top**) Four-class problem; (top-left) mean F value for the four classes *vs.* the number of samples per class for a 10% supervision ratio; (top-right) results with 50% of supervision; (**Bottom**) 12-class problem; (bottom-left) mean F value for the 12 classes for a 10% supervision ratio; (bottom-right) results with 50% supervision.

tion rate can be poor (as low as an F value of 0.5 or 0.3 for the four- and 12-class problems, respectively). However, after some number of samples (300 in our experiments), increasing the number of samples does not improve the results significantly. The only way to increase the classification rate is by introducing more supervision. In other words, as was expected, the supervision factor helps to obtain better F values, since more samples are used exclusively for training the mixture model of the corresponding class. Therefore, in a real application, the first goal is to achieve a minimum number of samples for every class so that the estimates can achieve a minimum quality. After that, if we can obtain more samples with known labels, we know that the performance will improve, since the problem is no longer about the amount of data, but the quality of the data.

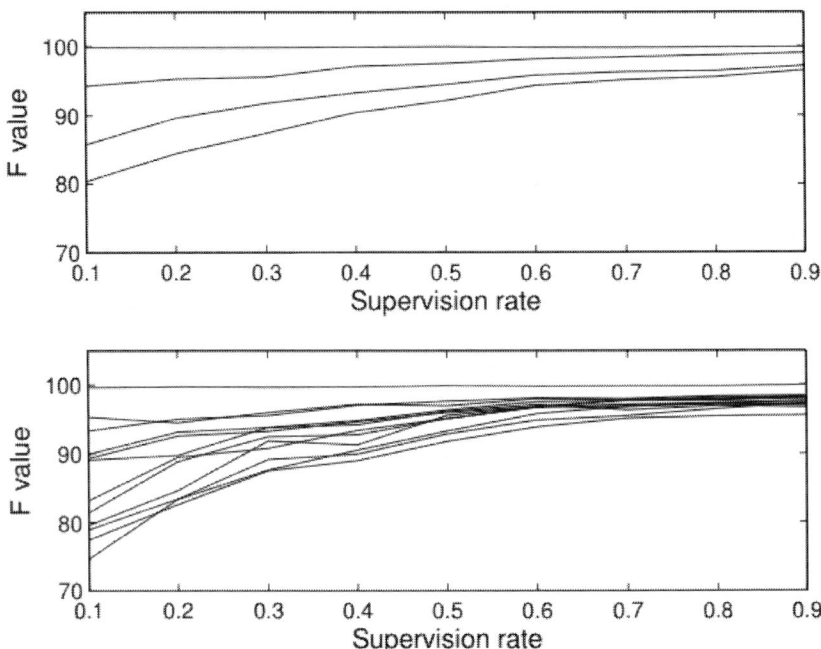

**Figure 7.** Influence of supervision. F value vs. percentage of supervision. (**Top**) Four-class problem; (**Bottom**) 12-class problem.

To clearly see the effect of the amount of supervision on performance, we show in Figure 7 the F value for different supervision rates: 0.1 (almost unsupervised), 0.3, 0.5, 0.7 and 0.9 (almost supervised). As was expected, the supervision allows one to model the distributions better, so the classification accuracy improves. Note that, even in the case of almost completely supervised classification, the classification is not perfect, since the classes are not separable, but the values are very close to a perfect classification, showing the good behavior of our algorithm

## 4.3. Results Depending on the Kind of Defect

Until now, we have studied the general performance of the algorithm, obtaining conclusions about the preprocessing of the feature vector (PCA dimensions reduction), the complexity of the model (how many Gaussians) and how the data size and supervision affects the results of the algorithm. Now, we will proceed to the analysis from the point of view of a single class.

In all previous figures, regardless if the 4-, 7- or 12-class problem was used, there was almost always a class that was perfectly classified even in inappropriate conditions (a feature vector with very few dimensions, bad class probability models with a small number of Gaussians and bad estimates of the model with few data and no supervision). This class was always the homogeneous material. It is quite logical to assume that a material that has no defect has a very different spectrum from defective materials. In a real implementation, we have no idea about the prior probabilities of the classes. We do not know if there are more defective than non-defective pieces or *vice versa*. Therefore, we have to remark again on the importance of using classification measures that are not sensitive to the prior probabilities of the classes; e.g., if we always decide that the piece is homogeneous and the prior of the non-defective material class is 0.95, we will obtain a 0.95 true positive rate, which can be misunderstood as the good performance of the algorithm if we only consider this measure.

The same could be said about the cost function. In a real implementation, attending to many other variables, such as the revenues, the owner of the system can change the decision making threshold in order to obtain the true positive/false alarm rate that is better for him; e.g., if many pieces are being classified as defective, he can change the decision rule that changes the cost function or, equivalently, assigning the piece to the defective class only when the posterior probability is higher than a given value instead of assigning it to the class with higher posterior probability.

In this section, we use nine Gaussians per class and 500 samples per class during training. The four-class problem essentially tests the ability of the algorithm to discriminate between defective and non-defective pieces. In Figure 8, we obtain the precision and recall values for each of the four classes (homogeneous, hole defect, crack defect and multiple defects) with respect to the supervision ratio, and in Table 2, we show the confusion matrix for the case of 0.1 supervision (the worst case).

As we can see in the figure, the results are very good. The precision of the homogeneous class is nearly 100% for any supervision value. In the case of low supervision, the precision is a little bit lower than 90% for the pieces with multiple defects and almost equal to 85% for the one defect blocks. However, the recall is almost perfect for the homogeneous and multiple defect classes in all supervision scenarios, *i.e.*, almost no piece in these classes is missed, regardless if they are included during the learning process. The lowest precision value of 75% is for the pieces with a crack and 0.1 supervision. Note that low supervision is the most realistic case for

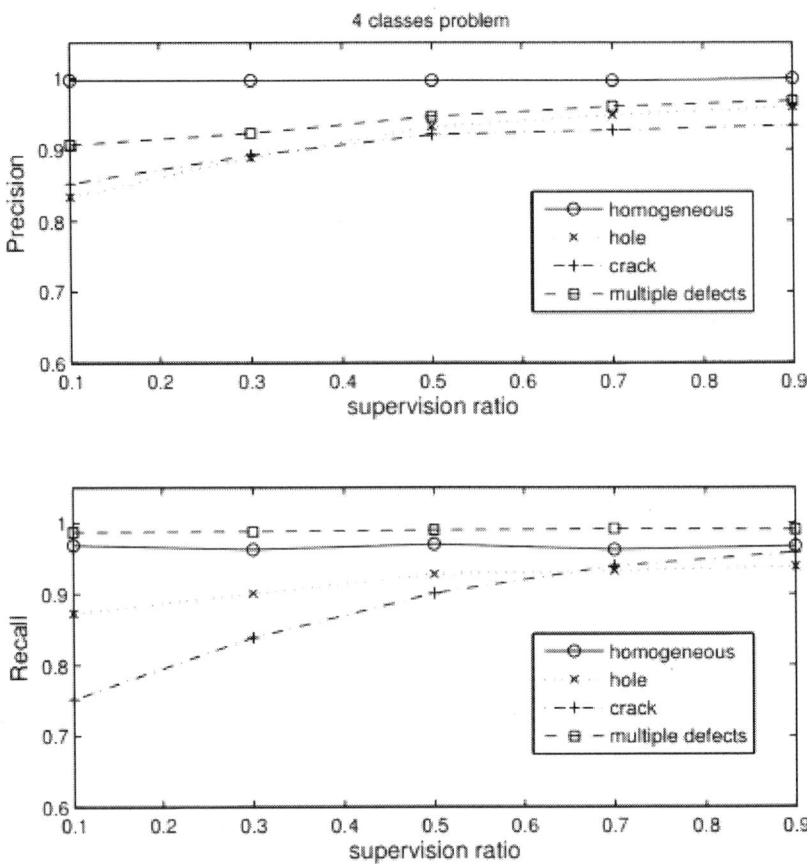

**Figure 8.** Precision and recall for the four-class problem vs. the supervision ratio.

**Table 2.** Confusion matrix for the 4-class problem (supervision 0.1).

|  | **Homogeneous** | **Hole** | **Crack** | **Multiple** |
| --- | --- | --- | --- | --- |
| homogeneous | 99.74 | 0.03 | 0.13 | 0.10 |
| hole | 0.28 | 83.05 | 15.99 | 0.68 |
| crack | 0.85 | 13.81 | 84.82 | 0.52 |
| multiple | 1.92 | 0.42 | 7.23 | 90.43 |

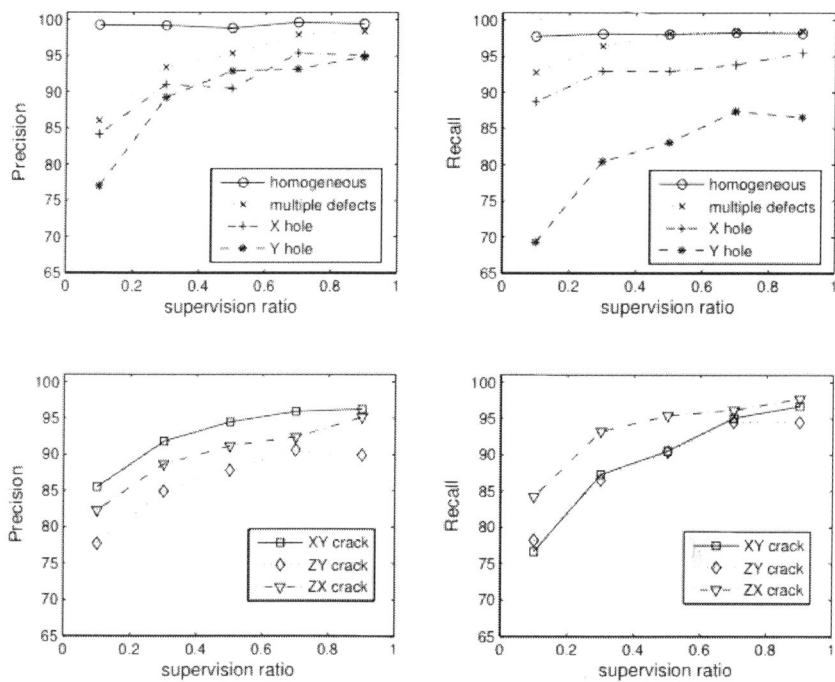

**Figure 9.** Precision and recall for the seven-class problem vs. supervision ratio.

industrial applications, since most defects are not observable and we want to retain the largest amount of pieces. Looking at Table 2, we see that most of the errors occur when one crack and hole classes are confused; *i.e.*, the cost would be drastically reduced if we consider that the kind of defect does not matter and we merge the hole and crack class into just one class.

In the seven-class problem, we split the hole and crack classes into subclasses, taking into account the direction of the defect: homogeneous, X hole, Y hole, XY crack, ZY crack, ZX crack and multiple defects. In Figure 9, we show the precision and recall values for the seven classes. For the sake of clarity, we split the results into different subplots.

Again, the homogeneous class is the best one, and increasing the supervision improves the performance for all classes. If we analyze the errors between classes, we find that most of the mistakes are between the Y hole and ZY crack classes. The values in the confusion matrix are shown in Table 3 only for the case of 0.5 supervision. However, the same occurs for the rest of the supervision values. A detailed understanding of this aspect

**Table 3.** Confusion matrix for the 7-class problem (supervision 0.5).

|          | Homog. | X Hole | Y Hole | XY Crack | ZY Crack | ZX Crack | Multiple |
|----------|--------|--------|--------|----------|----------|----------|----------|
| homog.   | 98.74  | 0.00   | 0.44   | 0.05     | 0.16     | 0.60     | 0.00     |
| X hole   | 1.05   | 89.95  | 5.58   | 0.42     | 1.68     | 1.21     | 0.11     |
| Y hole   | 0.06   | 1.76   | 92.55  | 0.30     | 4.72     | 0.61     | 0.00     |
| XY crack | 0.17   | 0.06   | 1.07   | 94.23    | 1.53     | 1.13     | 1.81     |
| ZY crack | 0.26   | 2.52   | 7.09   | 1.63     | 87.35    | 1.10     | 0.05     |
| ZX crack | 0.36   | 2.69   | 2.23   | 1.92     | 1.87     | 90.93    | 0.00     |
| multiple | 0.00   | 0.00   | 0.05   | 4.85     | 0.00     | 0.00     | 95.10    |

will require an in-depth analysis from the perspective of wave propagation, which is outside of the scope of this work. However, we may conjecture that due to the orientation of the impact and the similar width of both the ZY crack and the Y hole in the X direction (see Figure 3), the respective cross-sections "seen" by the impact point are similar, thus facilitating the confusion of these two kinds of defects.

In the 12-class problem, we introduce another distinction between defects (passing through or not passing through): homogeneous, through X hole, non-through X hole, through Y hole, non-through Y hole, through XY crack, non-through XY crack, through ZY crack, non-through ZY crack, through ZX crack, non-through ZX crack and multiple defects. In Figure 10, we show the precision and recall values for each of the twelve classes with respect to the supervision ratio. For the sake of clarity, we split the results into different figures, each one corresponding to four out of the 12 classes. Since the problem is much more complex, the results are good, but worse than in the previous scenarios. The detailed analysis of the confusion matrix (not shown for the sake of readability) reveals the same conclusion as in the seven-class problem: most mistakes are between holes and cracks that are aligned in the direction of the impact hammer.

### 4.4. ROC Curves

Since the algorithm assigns *a posteriori* probability for every sample, we can use this probability value as a score to obtain an ROC curve for every class. ROC curves have two important characteristics. First, they are insensitive to class probabilities. Second, they provide information about detection rate *vs.* false alarm rate, so a user can choose the working point of the

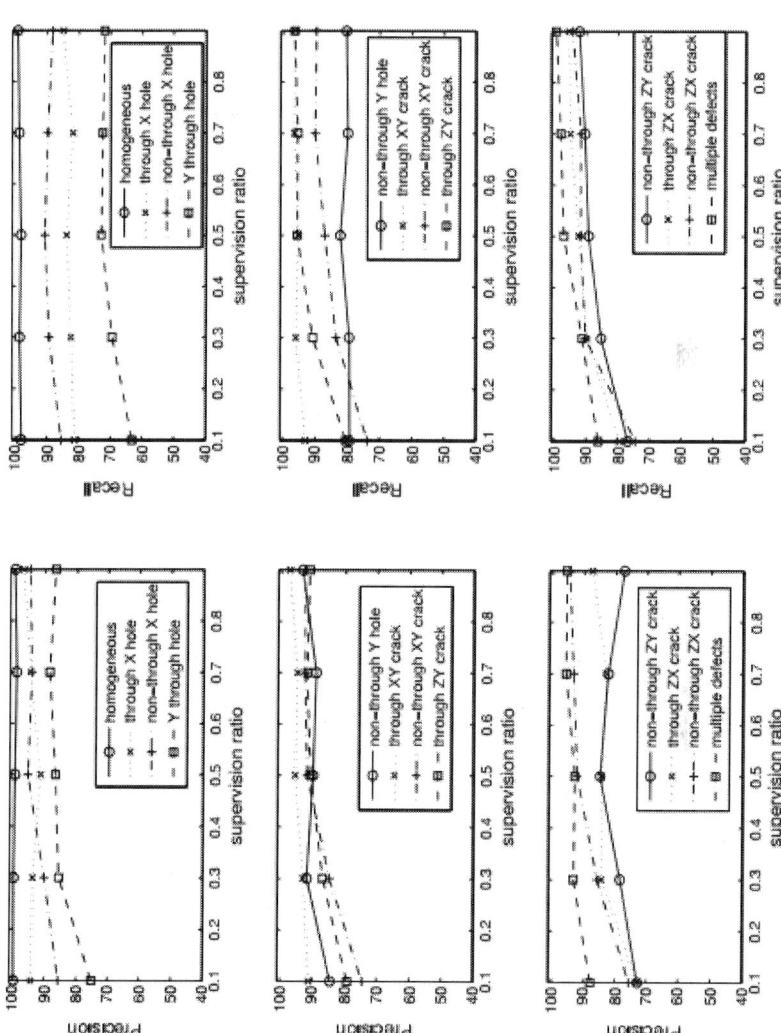

**Figure 10.** Precision and recall for the 12-classes problem vs. supervision ratio.

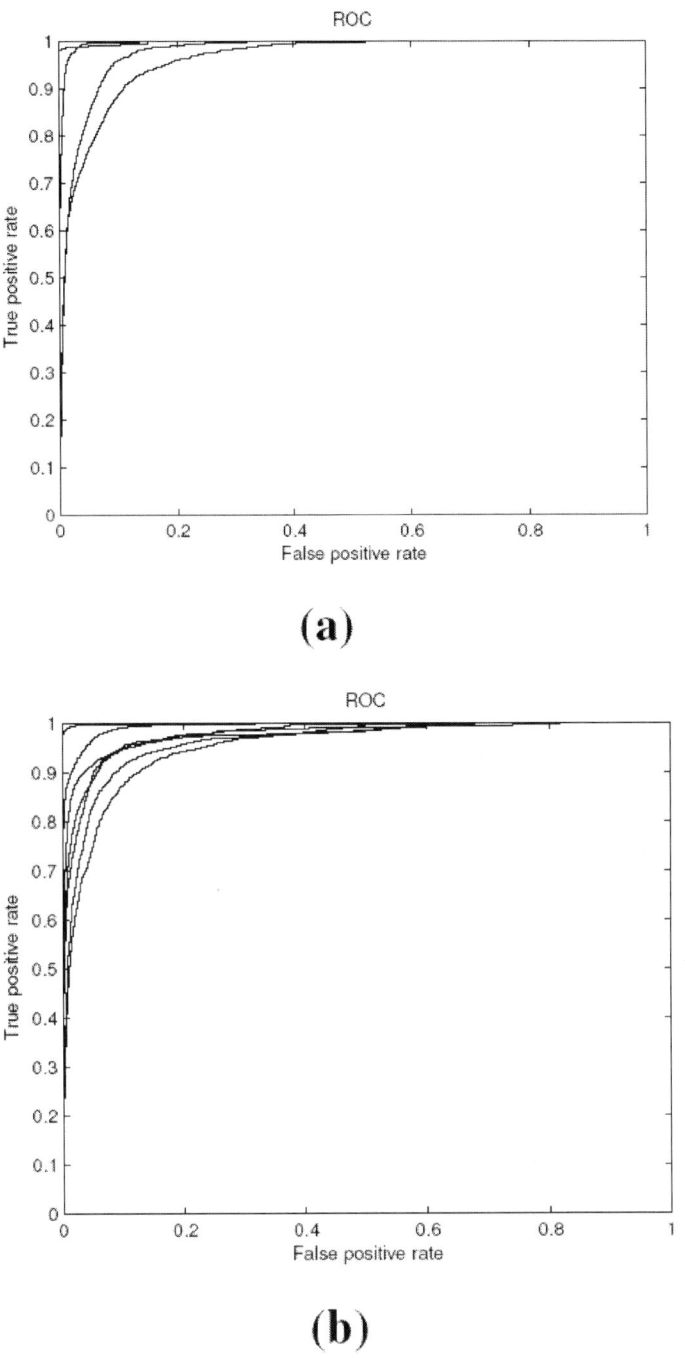

**Figure 11.** ROC curves. Detection rate vs. false alarm rate. (**a**) Four classes; (**b**) seven classes; (**c**) 12 classes.

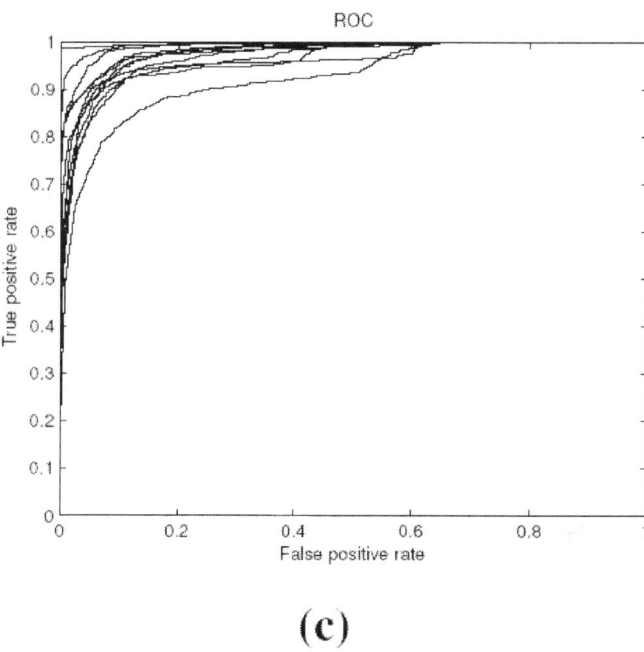

(c)

**Figure 11.** (Continued)

detector based on his preferences. Therefore, by analyzing the ROC graph, the user can choose the operating point (the pair true positive-false positive rates) according to his interest, e.g., a conservative detection system where we want to avoid the false positives, or just the point where the system obtains the best performance. According to this policy, we only have to adapt the threshold (score or probability) to the corresponding value. This is an important advantage of our Bayesian classifier with respect to other classification algorithms that only provide a binary decision, so it is difficult to evaluate the results beyond a misclassification point of view. Since we are working with up to twelve classes, it is important to know if the posterior distributions are similar or not, as this can give us an idea about the robustness of the classifier when the samples, priors or cost function change.

In Figure 11, we show the ROC curves for the 4, 7 and 12 classes for the 0.1 supervision case. As we can see, the graphs are clearly over the diagonal line in all cases, exhibiting the good performance of the algorithm. With this information, the user can select the operating point without having to learn or adapt any parameter in the system; just changing the detection threshold results in the false alarm rate that he wants.

## 5. DISCUSSION AND CONCLUSIONS

The statement of the problem from a statistical point of view permits the use of a generative model that is helpful in the understanding of the problem. Since we obtain the posterior probability for every class, we can use it for simple classification or for a more sophisticated analysis based on ROC curves.

In an industrial application, such as in the marble industry, flexibility is a key factor, since the user needs an adaptable tool where the classifier can be modified easily by considering economic reasons. However, this flexibility is more helpful if the user can learn from it. In other words, we need to quantify the change in the results not only in terms of classification-misclassification rates, as typical discriminative classifiers do. This is an advantage of our algorithm, since the user can analyze the posterior probabilities and learn from them in order to define the classification rule that is appropriate for them.

We have used the MoG model to approximate the conditional distribution of the data given a class. We have explained how to estimate the number of Gaussians and the rest of the parameters in order to obtain good performance while avoiding overfitting. Of course, the final results will depend on the separability of the classes.

A very good characteristic of the proposed algorithm is that it does not require a large amount of samples to estimate the parameters properly. In addition, we have shown how the quantity and quality of the data, *i.e.*, the supervision ratio, complement each other. This means that in a real application, we do not require the destruction of many pieces to obtain enough data, and, therefore, the implementation is affordable.

The results show that the homogeneous class is easily separable from the rest of the classes. An interesting conclusion is that homogeneous and defective specimens are rarely misclassified. When the number of classes increases, *i.e.*, when the defective pieces are subdivided into different subclasses, the problem is not as easy. This is especially true for the specimens that have similar defects, e.g., a hole and a crack with the same orientation. In this case, the only way to improve the performance is to better estimate the model parameters by increasing the supervision ratio. The price to pay is the economic cost, as we need to destroy more specimens to determine to which class they belong.

In order to help those who want to replicate our experiments, we give some advice. First, run multiple impact-echo experiments and discard the first ones. Second, the size of the test specimen plays an important role in the signals. It is necessary to perform the impact-echo test at several points on the surface to identify possible geometrical effects. In summary, it is advisable to test first the variables that can affect the results and to be sure that, in case you cannot control them, at least you can reduce their effects by running the experiments under different conditions, so that the effects are averaged and the signals are not biased.

## ACKNOWLEDGEMENTS

This work has been supported by Generalitat Valenciana under Grants PROMETEO II/2014/032, ISIC/2012/006 and GV/2014/034.

## AUTHOR CONTRIBUTIONS

Jorge Igual provided the method and the results. Addisson Salazar conceived of and performed the experiments and data. Gonzalo Safont helped in the collection of data. Luis Vergara supervised the theoretical statement of the problem. All authors revised the paper for intellectual content.

## CONFLICTS OF INTEREST

The authors declare no conflict of interest.

## REFERENCES

1. Hola, J.; Bien, J.; Sadowski, L.; Schabowicz, K. Non-destructive and semi-destructive diagnostics of concrete structures in assessment of their durability. *Bull. Pol. Acad. Sci. Tech. Sci.* **2015**, *63*, 87–96.
2. Chapelle, O.; Schölkopf, B.; Zien, A. *Semi-Supervised Learning*; Adaptive Computation and Machine Learning, MIT Press: Cambridge, MA, USA, 2006; p. 508.
3. Sansalone, M.J.; Streett, W.B. *Impact-Echo: Nondestructive Evaluation of Concrete and Masonry*; Bullbrier Press: Jersey Shore, PA, USA, 1997.

4. Jolliffe, I. *Principal Component Analysis*; Springer Verlag: New York, NY, USA, 1986.
5. YE, J.; Iwata, M.; Takumi, K.; Murakawa, M.; Tetsuya, H.; Kubota, Y.; Yui, T.; Mori, K. Statistical Impact-Echo Analysis Based on Grassmannn Manifold Learning: Its Preliminary Results for Concrete Condition Assessment. Proceedings of 7th European Workshop on Structural Health Monitoring, Nantes, France, 8–11 July 2014; pp. 1349–1356.
6. Popovics, J.; Song, W.; Achenbach, J.; Lee, J.; Andre, R. One-sided stress wave velocity measurement in concrete. *J. Eng. Mech.* **1998**, *124*, 1346–1353.
7. Gibson, A.; Popovics, J.S. Lamb wave basis for impact-echo method analysis. *J. Eng. Mech.* **2005**, *131*, 438–443.
8. Zhang, R.; Seibi, A.C. Impact-Echo Nondestructive Testing and Evaluation with Hilbert-Huang Transform. *Int. J. Mech.* **2010**, *4*, 105–112.
9. Abraham, O.; Leonard, C.; Cote, P.; Piwakowski, B. Time frequency analysis of impact-echo signals: numerical modeling and experimental validation. *ACI Mate. J.* **2000**, *97*, 645–657.
10. Baggens, O.; Ryden, N. Systematic errors in Impact-Echo thickness estimation due to near field effects. *NDT&E Int.* **2015**, *69*, 16–27.
11. Hola, J.; Sadowski, L.; Schabowicz, K. Nondestructive identification of delaminations in concrete floor toppings with acoustic methods. *Autom. Constr.* **2011**, *20*, 799–807.
12. Völker, C.; Shokouhi, P. Multi sensor data fusion approach for automatic Honeycomb detection in concrete. *NDT&E Int.* **2015**, *71*, 54–60.
13. Cho, Y.S.; Hong, S.U.; Lee, M.S. Multi sensor data fusion approach for automatic Honeycomb detection in concrete. *NDT&E Int.* **2009**, *24*, 277–288.
14. Zhou, X.; Wang, Z.; Yan, B. Nondestructive testing method of grouting quality for prestressed pipe. *China J. Highw. Transp.* **2011**, *24*, 64–71.
15. Sadowski, L.; Hoła, J. Neural prediction of the pull-off adhesion of the concrete layers in floors on the basis of nondestructive tests. *Procedia Eng.* **2013**, *57*, 986–995.

16. Azari, H.; Nazarian, S.; Yuan, D. Assessing sensitivity of impact echo and ultrasonic surface waves methods for nondestructive evaluation of concrete structures. *Constr. Build. Mater.* **2014**, *71*, 384–391.
17. Vergara, L.; Gosálbez, J.; Fuente, J.; Miralles, R.; Bosch, I.; Salazar, A.; López, A.; Domínguez, L. Ultrasonic nondestructive testing on marble rock blocks. *Mater. Eval.* **2004**, *62*, 73–78.
18. Shokouhi, P.; WÃűstmann, J.; Schneider, G.; Milmann, B.; Taffe, A.; Wiggenhauser, H. Nondestructive detection of delamination in concrete slabs. *Transp. Res. Rec.* **2011**, *2551*, 103–113.
19. Dempster, A.P.; Laird, N.M.; Rubin, D.B. Maximum likelihood from incomplete data via the EM algorithm. *J. R. Stat. Soc.* **1977**, *39*, 1–38.
20. Chiang, C.; Cheng, C. Detecting rebars and tubes inside concrete slabs using continuous wavelet transform of elastic waves. *J. Mech.* **2004**, *20*, 297–302.
21. Yeh, P.L.; Liu, P.L. Application of the wavelet transform and the enhanced Fourier spectrum in the impact echo test. *NDT&E Int.* **2008**, *41*, 382–394.
22. Chaudhary, M.T.A. Effectiveness of Impact Echo testing in detecting flaws in prestressed concrete slabs. *Constr. Build. Mater.* **2013**, *47*, 753–759.
23. Carino, N.J.; ASCE Member. The impact-echo method: An overview. Proceedings of the 2001 Structures Congress & Exposition, Washington, DC, USA, 21–23 May 2001; pp. 21–23.
24. Salazar, A.; Vergara, L.; Llinares, R. Learning material defect patterns by separating mixtures of independent component analyzers from NDT sonic signals. *Mech. Syst. Signal Process.* **2010**, *24*, 1870–1886.
25. Learned-Miller, E.G.; Fisher, J.W., III. ICA Using Spacings Estimates of Entropy. *J. Mach. Learn. Res.* **2003**, *4*, 1271–1295.
26. Fawcett, T. An Introduction to ROC Analysis. *Pattern Recogn. Lett.* **2006**, *27*, 861–874.

# CHAPTER 7

## OPTIMAL ALLOCATION OF RADIO RESOURCE IN CELLULAR LTE DOWNLINK BASED ON TRUNCATED DYNAMIC PROGRAMMING UNDER UNCERTAINTY

Abayomi M. Ajofoyinbo, Kehinde O. Orolu

Department of Systems Engineering, Faculty of Engineering Complex, University of Lagos, Lagos, Nigeria

## ABSTRACT

In the Cellular Long-Term Evolution (LTE) downlink, the smallest radio resource unit a Scheduler can assign to a user is a Resource Block (RB). Each RB consists of twelve (12) adjacent Orthogonal Frequency Division Multiplexing (OFDM) sub-carriers with inter-subcarrier spacing of 15 kHz. Over the years, researchers have investigated the problem of radio resource allocation in cellular LTE downlink and have made useful contributions. In an earlier paper for example, we proposed a deterministic dynamic programming based technique for optimal allocation of RBs in the downlink of multiuser Cellular LTE System. We found that this proposed methodology optimally allocates RBs to users at every transmission instant, but the computational time associated with the allocation policy was high. In the current work, we propose a truncated dynamic programming based technique for efficient and optimal allocation of radio resource. This paper also addresses uncertainty emanating from users' mobility within a Cell coverage area. The objective is to significantly reduce the computational time and dynamically select applicable modulation scheme (i.e., QPSK, 16QAM, or 64QAM) in response to users' mobility. We compare the proposed scheme with the Fair allocation and the earlier proposed dynamic programming based techniques. It is shown that the proposed methodology is more efficient in allocating radio resource and has better performance than both the Fair Allocation and the deterministic dynamic programming based techniques.

## KEYWORDS

Optimal Allocation; Resource Block; Truncated; Dynamic Programming; Uncertainty

## 1. INTRODUCTION

In the 3rd Generation Partnership Project (3GPP) Long Term Evolution (LTE), Orthogonal Frequency Division Multiplexing (OFDM) is chosen as the radio transmission scheme. Packet scheduling is one of the LTE Radio Resource Management (RRM) functions, which is responsible for allocating resources to users. In taking scheduling decisions, RRM takes into account the Channel Quality Information (CQI) from the User Equipment (UE), the Quality of Service (QoS) requirements and the buffer status. The downlink of LTE Cellular System is based on Orthogonal Frequency Division Multiple Access (OFDMA), which provides efficient multi-user access and intra-cell interference avoidance. The smallest radio resource unit that the scheduler can assign to a user is a Resource Block (RB) [1]. In the downlink of LTE Cellular System, radio resources are partitioned in both the frequency and time domains. Access to radio resource is controlled in terms of frames and frequency channels. These frames and frequency channels are referred to as the OFDM Resource Blocks (RBs). Each RB consists of 12 adjacent OFDM sub-carriers with intersubcarrier spacing of 15 kHz. Each sub-frame (i.e., 1 ms in the time domain) consisting of two (2) RBs is divided into fourteen (14) symbols of which up to three symbols at the start of the sub-frame can be used for controlling channel signalling. The allocation of RBs to user is done by the scheduler and scheduling decision is taken during each Transmission Time Interval (TTI) of 1 ms for each sub-frame. The allocated RBs and the selected modulation and coding schemes are signalled to the scheduled user on the Physical Downlink Control Channel

(PDCCH) (Iosif and Banica [2]). The dynamic scheduler also interact with the Hybrid Automatic Repeat Request (HARQ) manager as it is responsible for scheduling retransmissions and it may also take into account the QoS attributes and buffer information (Holma and Toskala [3]). In recent years, researchers have investigated problem of radio resource allocation in LTE Downlink and have made useful contributions. Huang et al. [4] investigated the problem of gradient-based scheduling and resource allocation for the downlink of a Cellular OFDM system, which reduces to solving a convex optimisation problem in each time slot. The work considered scheduling and resource allocation for the downlink of a Cellular OFDM system, with various practical considerations including integer tone allocations, different sub-channelization schemes, maximum signal-to-noise ratio constraint per tone, and "self-noise" due to channel estimation errors and phase noise. The authors proposed an algorithm that automatically yields an integer carrier allocation. Kasier and Ahmed [5] proposed a priority based resource allocation algorithm for heterogeneous services in the relay enhanced Orthogonal Frequency Division Multiple Access (OFDMA) downlink systems. The aim of the research was to maximise the system throughput while satisfying the Quality of Service (QoS) of the heterogeneous services comprising Real Time (RT) and NonReal-Time (NRT) services. The proposed algorithm reduces the outage probability of the system and increases the system throughput. Anding et al. [6] presented an adaptive sub-carrier and power allocation scheme for OFDMA systems according to their different QoS requirement(s) and traffic type. The proposed algorithm maximised the transmission data rate while satisfying total power constraint and a certain Bit Error Rate (BER) requirement. This algorithm first allocates sub-carriers and bits to high priority users according to their rates and Bit Error Rate (BER) requirement, and then distributes the residual resources to the ordinary users in terms of the proportional fairness principle. Simulation results show that the proposed method has better performance than the existing algorithms. In [7], Kumar et al. proposed new heuristic algorithms to solve the sub-carrier, bit, and power allocation in polynomial computational complexity. The two-stage allocation algorithms include the initial sub-carrier assignment and the iterative improvement two-steps. Simulation results show that the performance of the proposed heuristic algorithm is close to that of the optimum solution. Li and Liu [8] proposed a two-level resource allocation scheme, where the first level coordinates Cells while the second level performs per-Cell optimisation. Moreover, Koutsopoulos and Tassiulas [9] presented two classes of centralised heuristic algorithms. The first one considers each sub-carrier in each Cell. The results obtained show that the first class of heuristics performs better and quantify the input of different parameters on system performance. Thonabalasingham et al. [10] worked on joint allocation of various radio resources and concluded that joint allocation of various radio resources has a clear potential over methodologies that allocate single resource. In [11], Koutsimanis and Fodor worked on the elastic nature of data applications. In this work, each user is associated with a minimum and maximum Resource Block requirements and the resource allocation problem consists of maximising the overall throughput such that these requirements are met. Wong and Evans [12] developed optimal resource allocation algorithms for OFDMA systems assuming the availability of only partial (imperfect) Channel State Information (CSI). The authors considered both continuous and discrete weighted sum rate maximisation subject to total power constraints, and average bit error rate constraints for the discrete rate case. The work of Hosein [13] focused on the optimal allocation of power and bandwidth with the objective of maximising the sector-wide throughput. The work determined the regions within which 1) a frequency reuse factor of unity is optimal; 2) orthogonal frequency allocation is optimal; and 3) joint processing of signals from both sectors is optimal. Zhou et al. [14] proposed a Genetic Algorithm (GA) based cross-layer resource allocation for the downlink multiuser wireless OFDM system with heterogeneous traffic. The GA was used to maximise the sum of weighted capacities of multiple traffic queues at the Physical layer, where the weights are determined by the Medium Access Control (MAC) layer. In [15], Ergen et al. proposed a fair scheduling scheme that allocates sub-carriers to users and then determine the number of bits transmitted on each subcarrier. In an earlier paper, Ajofoyinbo and Orolu [16] proposed a dynamic programming based technique for optimal allocation of radio resource in the downlink of multi-user LTE Cellular System. The main contribution of this work is optimal allocation of RBs to users. Indeed, this optimal allocation of RBs yields maximum cumulative contribution to the overall network throughput at every transmission instant, but the associated computational time is normally high. In the current work, a truncated dynamic programming based technique is proposed for efficient and optimal allocation of radio resource in the downlink of a multi-user LTE Cellular System. The methodology also addresses uncertainty emanating from users' mobility within a Cell coverage area.

The proposed technique maximises instantaneous channel throughput by efficiently and optimally allocating RBs to users, using less computational time.

The remainder of the paper is organized as follows. Problem formulation is presented in Section 2. This includes systems modelling and allocation policy. The problem solution and simulation of the proposed model is presented in Section 3. This is followed by a discussion of simulation results in Section 4. Section 5 concludes the paper.

This paper contributes to knowledge in the field of optimal allocation of RBs to achieve maximum instantaneous throughput in the downlink of multi-user LTE Cellular System.

## 2. PROBLEM FORMULATION

In this section, we present the systems modelling and allocation policy for the proposed Truncated Dynamic Programming (TDP) based technique.

### 2.1. Systems Modelling for the TDP Technique

The proposed TDP under uncertainty model consists of fifty-one (51) states, namely: 0, 2, 4, ..., 98, 100. These states represent the number of allocated RBs. The RBs are allocated in multiples of 2 RBs, starting with zero RB. We note that in the earlier proposed deterministic dynamic programming based technique; the system consisted of six (6) states. Moreover, the proposed model consists of n stages, which represent the scheduled users on the LTE downlink. In developing the recursive optimisation procedure, we obtain a solution of the overall nstage problem by first solving a one-stage problem, and by sequentially including one stage at a time, solving one-stage problems until the overall optimum is reached. The applicable dynamic programming procedure is the Back Induction (BI) process wherein the first stage to be analysed is the final stage of the problem and the n-stage problem is solved moving backwards one stage at a time until all stages are included to obtain optimal solution (Hillier and Lieberman [17]). We define the following variables:

$x_k$ = Possible allocation of RBs

$p_j(x_k)$ = Throughput obtainable from possible allocation of $x_k$ RBs to user j.

$N_{os}^{j}$ = Number of OFDM symbols for user j.

$N_{mod}^{j}$ = Number of bits per OFDM Symbol for user j.

$N_{RE}$ = Number of Resource Elements (RE).

$m$ = Total number of RBs.

$n$ = Total number of users.

$s$ = Number of RBs still available for allocation.

The objective is to choose $x_k$ so as to

$$\text{Maximize} \quad Z = \sum_{j=1}^{n} p_j(x_k) \tag{1}$$

$$\text{Subject to} \quad \sum_{k=0}^{50} x_k \leq m \tag{2}$$

Where

$$k = 0, 1, 2, \cdots, 50 \tag{3}$$

$$x_k = 0, 2, 4, \cdots, 100 \tag{4}$$

$$j = 1, 2, \cdots, n \tag{5}$$

We note that

$$p_j(x_k) = 12 \times N_{os}^{j} \times N_{mod}^{j} \times x_k \tag{6}$$

and

$$\sum_{k=0}^{50} x_k \leq 100 \qquad (7)$$

## 2.2. Allocation Policy and Recursion for the Proposed Truncated Dynamic Programming (TDP) Technique

WE NOTE THAT THE TRUNCATED DYNAMIC PROGRAMMING analysis starts from the last user (i.e., j = n, the final stage of the problem), but actual allocation of RBs starts with the first user (i.e., j = 1).

In this proposed model, the computations leading to the allocation of RBs follow a recursive procedure. We define $f_j(s, x_k)$ as the cumulative contribution of user $j$ to the objective function, given $s$ available RBs and $x_k$ allocated RBs to user $j$.

For the last user (i.e., $j = n$),

$$f_{j+1}^*(s - x_k) = 0 \qquad (8)$$

$$f_j(s, x_k) = p_j(x_k) \qquad (9)$$

From the last-but-one-user (i.e., j = n – 1) to the first user (i.e., j = 1), we apply Equation (10).

$$f_j(s, x_k) = p_j(x_k) + f_{j+1}^*(s - x_k) \qquad (10)$$

## 2.3. Motivation for Truncating the Dynamic Programming (DP) Procedure

The motivation for truncating the dynamic programming procedure is the determination of the point, beyond which it is no longer efficient to allocate RBs to a particular user given available RBs. To achieve this, we compare current computed cumulative throughput with the last two consecutive preceding cumulative throughputs. If the current cumulative throughput is less than or equal to the last two consecutive preceding cumulative throughputs, then the procedure is terminated (or truncated). The decision is to choose the maximum cumulative contribution to total channel throughput among computed cumulative contributions (or throughputs).

In computing throughput of individual user given available RBs, we invoke Equation (10) and execute the following truncated dynamic programming procedure. In the course of executing this procedure, the truncation conditions are checked before every computation of cumulative contribution (or throughput).

**Computation of Cumulative Throughput and Truncation Procedure**
For j = 19, 18, ..., 1 BEGIN
For k = 0, 1, 2, ..., 50 BEGIN

$$f_j(s, x_k) = p_j(x_k) + f^*_{j+1}(s - x_k) \tag{11}$$

IF $\left( f_j(s, x_k) \leq f_j(s, x_{k-1}) \right)$ and $\left( f_j(s, x_{k-1}) \leq f_j(s, x_{k-2}) \right)$

THEN Truncate ELSE Compute cumulative throughput for next $x_k$
END END The optimal allocation decision for user j, in relation to s and $x_k$, is the maximum of the cumulative contributions to the total throughput. This is given by:

$$f^*_j(s) = \max_{x_k = 0, 2, \cdots, s} \left\{ f_j(s, x_k) \right\} \tag{12}$$

Thus, the corresponding allocated RBs is given by:

$$x_k^* = \arg\left(\max_{x_k=0,2,\cdots,s}\{f_j(s,x_k)\}\right) \qquad (13)$$

## 2.4. Effect of Truncation on Recursion

Without truncation, the back induction dynamic programming process computes cumulative contributions for all possible allocation of RBs (i.e., $x_k$), given available RBs (i.e., s). The Scheduler takes decision on optimal allocation of RBs at the end of all computations. With the truncation procedure however, the recursion is terminated when the truncation conditions are met; thereby reducing computational time and usage of other computational resources. Thus, in terms of contribution to the total network throughput, the truncation exercise enables the Scheduler to determine the point beyond which it is no longer efficient to allocate RBs to users much earlier than without truncation. The advantage to the network is that the time and other resources that would have been committed to computations beyond the truncation point are saved. In LTE, new scheduling decisions are taken in every Transmission Time Interval (TTI), which is 1 ms. The reduced computational time and timely decision on allocation of RBs would have positive effects on time of scheduling decisions.

## 3. PROBLEM SOLUTION AND SIMULATION RESULTS

The simulation and users' transmission parameters are presented in Tables 1 and 2 respectively. At every scheduling instant, three modulation schemes are available to every user, namely: QPSK, 16QAM and 64QAM. The applicable modulation scheme depends on user's location within a Cell coverage area. Uncertainty regarding user's location is encapsulated in the truncated dynamic programming model at this stage. For example, a user with good signal strength may start transmission using 64QAM, and subsequently change at later transmission instant(s) to QPSK (or 16QAM) modulation scheme in response to change in location. The proposed truncated dynamic programming based technique was simulated on a Personal Computer us-

**Table 1.** Simulation parameters.

| | |
|---|---|
| Channel bandwidth | 20 MHz |
| Total number of RBs | 100 |
| Number of subcarriers per RBs | 12 |
| Modulation technique (bits/OFDM symbol) | QPSK(2), 16QAM(4), 6AQAM(6) |
| Minimum number of RBs to a user | 0 |
| Maximum number of RB to a user | 100 |
| Number of users ($i.e., n$) | 20 |
| Number of states in the truncated dynamic programming model | 51 ($i.e., 0, 2, 4, \cdots, 98, 100$) |
| $x_k$ | 0, 2, 4, 6, $\cdots$, 98, 100 |
| $S$ | 0, 2, 4, 6, $\cdots$, 98, 100 |
| OFDM symbols per slot (RB) | 7 |

**Table 2.** Users' transmission parameters.

| User | Data size (bits) | No of bits per OFDM symbol (N_mod): (QPSK = 2; 16QAM = 4; 64QAM = 6) | No. of OFDM symbols (N_os) per 2 RBs | Resource elements (12 × N_os) | Total bits per subframe (2 RBs) |
|---|---|---|---|---|---|
| 1 | 2000 | 6 | 11 | 132 | 792 |
|   |      | 4 | 9  | 108 | 432 |
|   |      | 2 | 7  | 84  | 168 |
| 2 | 4000 | 6 | 11 | 132 | 792 |
|   |      | 4 | 9  | 108 | 432 |
|   |      | 2 | 7  | 84  | 168 |
| 3 | 6000 | 6 | 11 | 132 | 792 |
|   |      | 4 | 9  | 108 | 432 |
|   |      | 2 | 7  | 84  | 168 |
| 4 | 8000 | 6 | 11 | 132 | 792 |
|   |      | 4 | 9  | 108 | 432 |
|   |      | 2 | 7  | 84  | 168 |
| 5 | 10000 | 6 | 11 | 132 | 792 |
|   |      | 4 | 9  | 108 | 432 |
|   |      | 2 | 7  | 84  | 168 |
| 6 | 2000 | 6 | 11 | 132 | 792 |
|   |      | 4 | 9  | 108 | 432 |
|   |      | 2 | 7  | 84  | 168 |
| 7 | 4000 | 6 | 11 | 132 | 792 |
|   |      | 4 | 9  | 108 | 432 |
|   |      | 2 | 7  | 84  | 168 |
| 8 | 6000 | 6 | 11 | 132 | 792 |
|   |      | 4 | 9  | 108 | 432 |
|   |      | 2 | 7  | 84  | 168 |
| 9 | 8000 | 6 | 11 | 132 | 792 |
|   |      | 4 | 9  | 108 | 432 |
|   |      | 2 | 7  | 84  | 168 |

| | | | | | |
|---|---|---|---|---|---|
| 10 | 10000 | 4 2 6 | 9 7 11 | 108 84 132 | 432 168 792 |
| 11 | 2000 | 2 4 6 | 7 9 11 | 84 108 132 | 168 432 792 |
| 12 | 4000 | 4 2 6 | 9 7 11 | 108 84 132 | 432 168 792 |
| 13 | 6000 | 2 4 6 | 7 9 11 | 84 108 132 | 168 432 792 |
| 14 | 8000 | 2 4 6 | 7 9 11 | 84 108 132 | 168 432 792 |
| 15 | 10000 | 2 4 6 | 7 9 11 | 84 108 132 | 168 432 792 |
| 16 | 2000 | 2 4 6 | 7 9 11 | 84 108 132 | 168 432 792 |
| 17 | 4000 | 2 4 6 | 7 9 11 | 84 108 132 | 168 432 792 |
| 18 | 6000 | 2 4 6 | 7 9 11 | 84 108 132 | 168 432 792 |
| 19 | 8000 | 2 4 6 | 7 9 11 | 84 108 132 | 168 432 792 |
| 20 | 10000 | 4 2 | 9 7 | 108 84 | 432 168 |

ing MATLAB Ver 7.10.0.449 (R2010a). Simulation results based on the proposed truncated and the existing dynamic programming techniques are obtained for the 20 network users. We are however constrained by limited space in this paper to present results for only the 18th user (i.e., j = 18) to demonstrate the efficiency and effectiveness of the proposed technique.

In the current paper, Option 1 represents data in Row 1 (i.e., from columns 3 to 6) for all users in **Table 2**. Similarly, Options 2 and 3 represent data in Rows 2 and 3 respectively (i.e., from Columns 3 to 6) for all users in **Table 2**. The management of uncertainty regarding dynamic selection of a modulation scheme shows the robustness of the proposed technique.

## 3.1. System Flowchart

The system flowchart for the implementation of the proposed technique is presented in **Figure 1**.

## 3.2. Simulation Results

Simulation results for the proposed Truncated Dynamic Programming (TDP) under uncertainty are presented in Figures 2, 4 and 6. We also present comparative results for the corresponding Dynamic Programming (DP) based technique in Figures 3, 5 and 7.

We further compare the performance of the proposed TDP technique with that of the Fair Allocation technique. The instantaneous cumulative throughputs for both techniques, based on Options 1, 2 and 3, are presented in Tables 3-5 respectively. The Fair allocation technique ensures fairness in allocation of resources to network users independent of traffic characteristics. The technique requires that each network user is allocated some amount of available radio resource for transmission.

To ensure realistic comparison and fairness, we allocate 6 RBs to users with good signal strength (i.e., users using 64QAM modulation scheme) and 4 RBs to users with poor signal strength (i.e., users using 16QAM or QPSK modulation scheme).

The graphs showing comparative cumulative throughputs for the TDP and Fair Allocation techniques are presented in **Figure 8**.

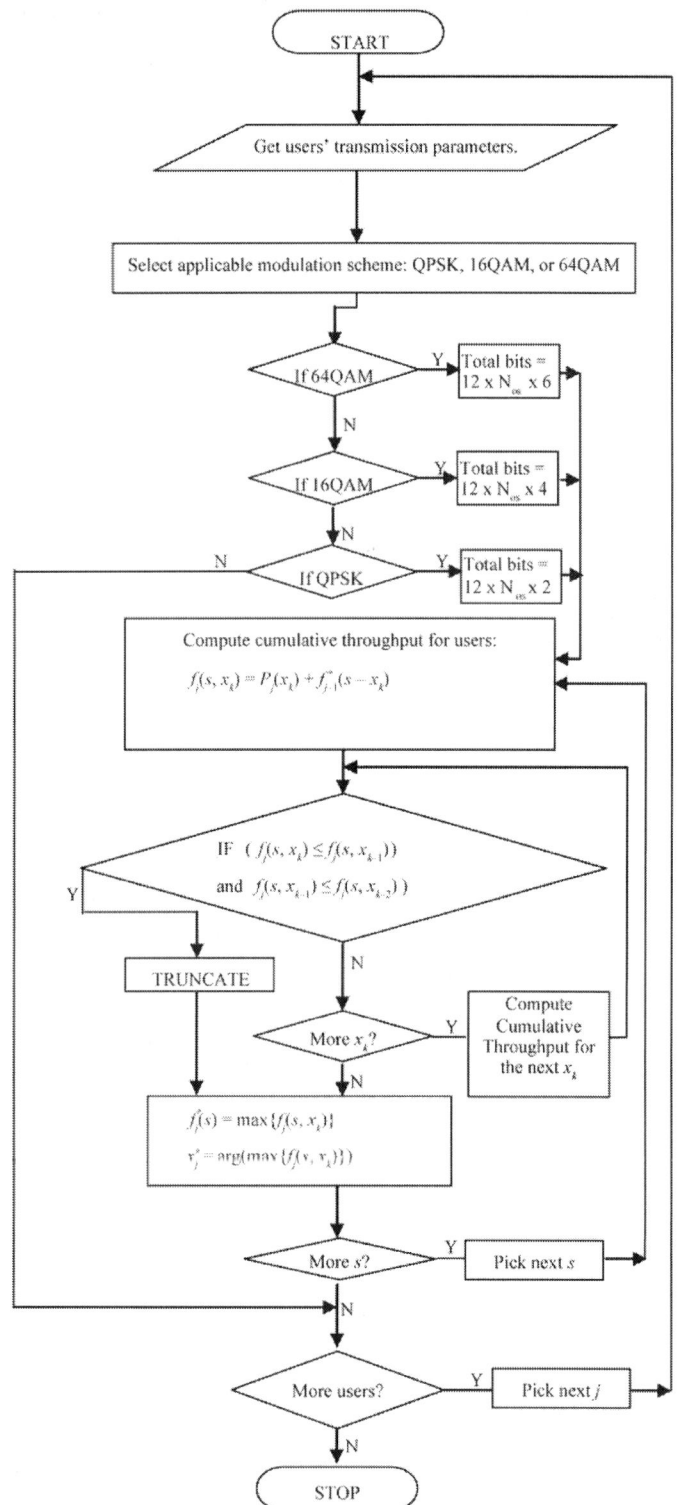

**Figure 1**. Systems flowchart.

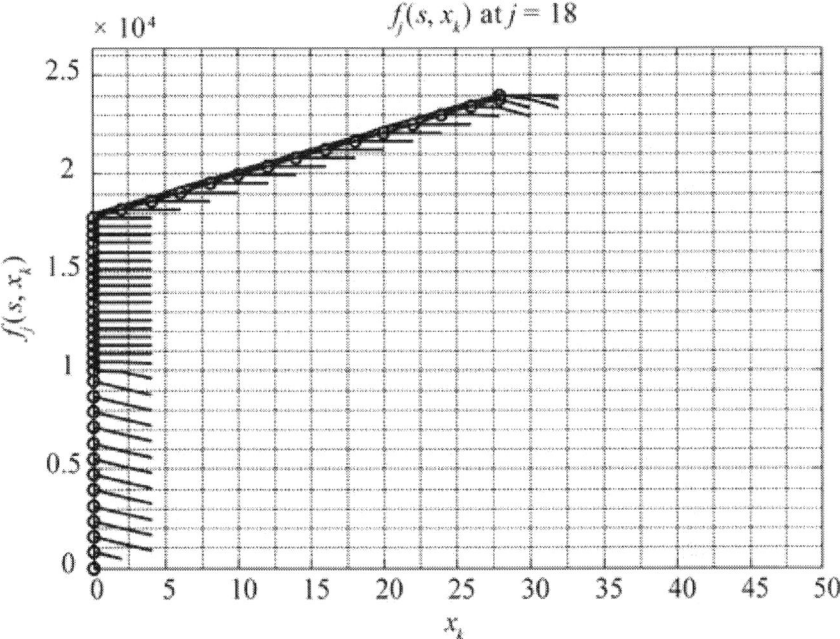

**Figure 2.** TDP—Option 1.

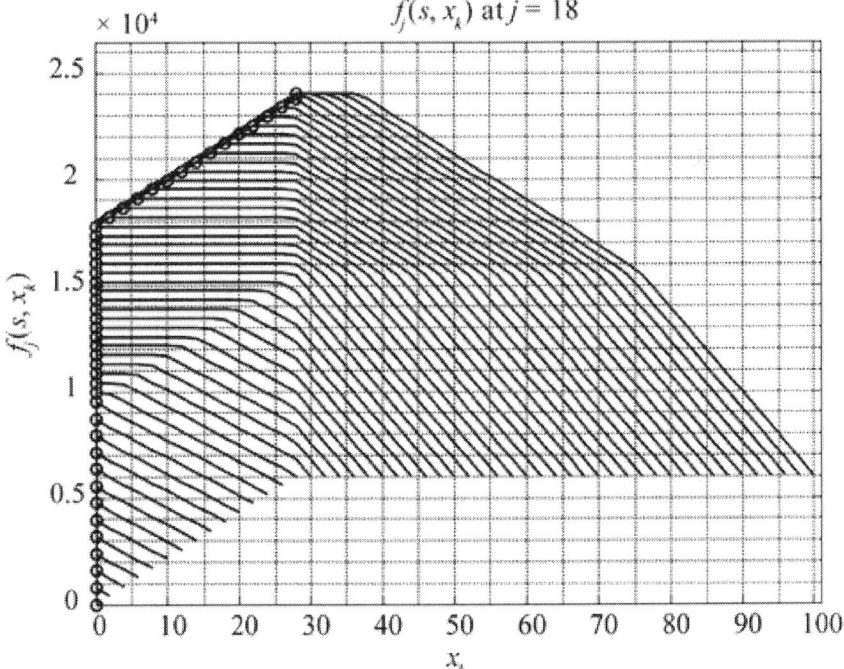

**Figure 3.** DP—Option 1.

180  Dynamic Programming and Bayesian Inference, Concepts and Applications

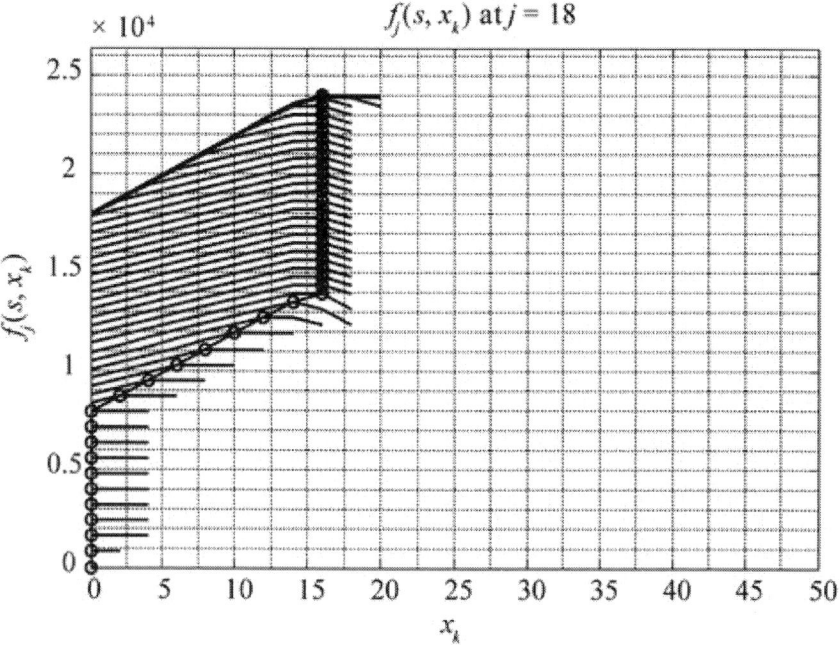

**Figure 4.** TDP—Option 2.

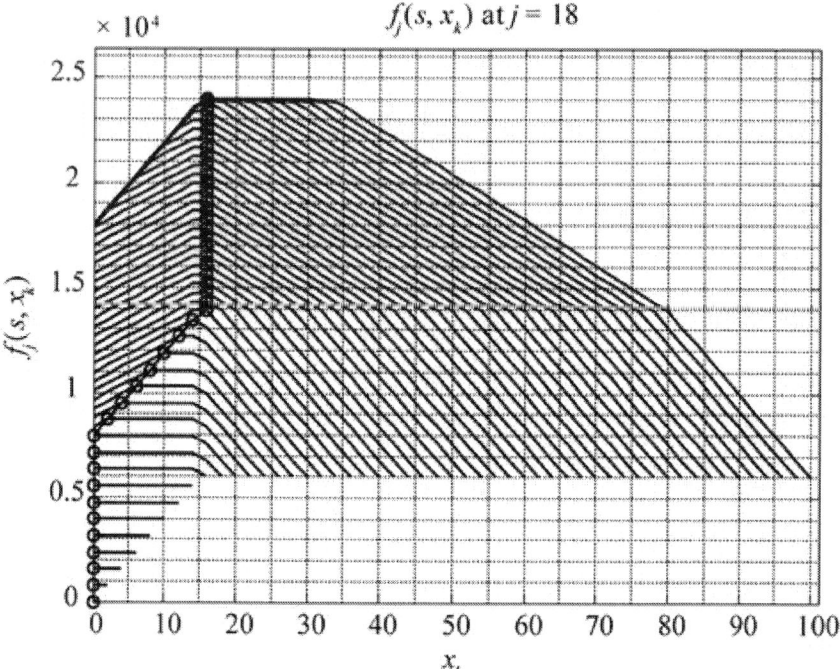

**Figure 5.** DP—Option 2.

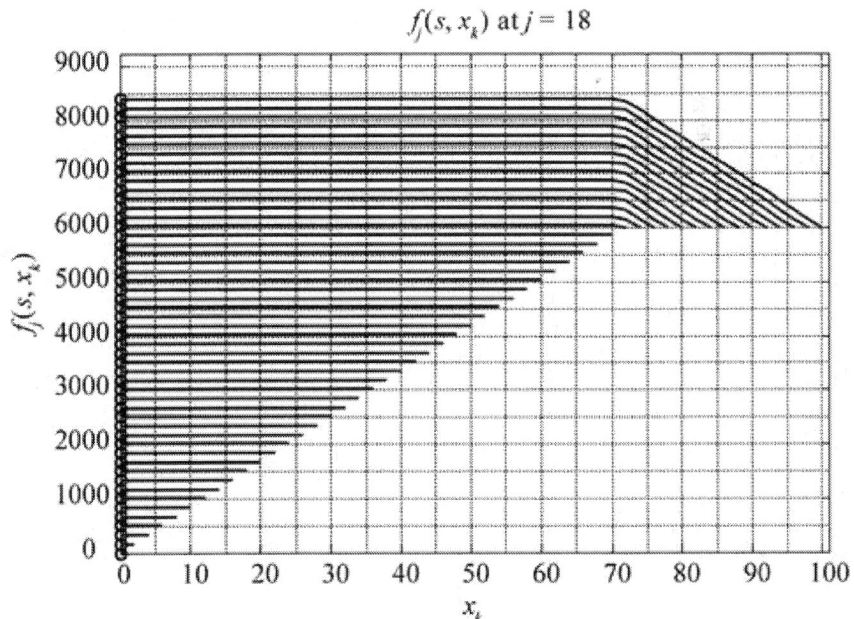

**Figure 6.** TDP—Option 3.

**Figure 7.** DP—Option 3.

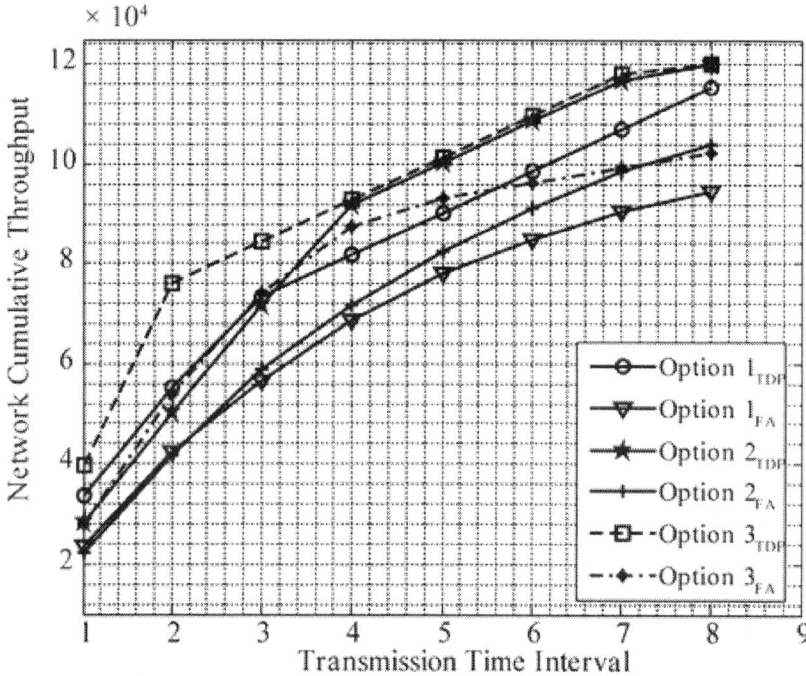

**Figure 8.** Network throughput for the Fair Allocation and Truncated Dynamic Programming based techniques.

## 4. DISCUSSION OF THE SIMULATION RESULTS

In the proposed truncated dynamic programming under uncertainty technique, the system can adapt dynamically to changing locations of users (corresponding to changing signal strength) as a result of users' mobility. The computations of instantaneous cumulative throughputs are based on the applicable modulation scheme (i.e., QPSK, 16QAM or 64QAM). Since the location of user within a Cell coverage area is uncertain, selection of applicable modulation scheme is also uncertain. Thus, the optimal decision at each stage can be taken only after the outcome of the uncertain event (i.e., selection of the applicable modulation scheme).

We note that for an optimal policy, whatever the current state and decision, the remaining decisions relating to the resulting state from the current decision must be an optimal policy. The optimality of the proposed allocation scheme is assured by the applicable dynamic programming procedure.

For the current work, the applicable dynamic programming procedure is the Back Induction (BI) process wherein the first stage to be analysed is the final stage of the problem and the n-stage problem is solved moving backwards one stage at a time until all stages are included to obtain optimal solution.

In this paper, we assume 20 network users within a Cell coverage area. Cells are geographical areas that are covered by radio signals from a Base Station. Allocated RBs ranges from 0 to 100, in multiples of 2 RBs (i.e., $x_k$: 0, 2, 4, 6, ..., 98, 100). In computing cumulative contributions using the proposed technique, the system compares current cumulative contribution with the last two preceding consecutive cumulative contributions. If the current computed cumulative contribution is less than or equal to the first preceding cumulative contribution and the first preceding cumulative contribution is less than or equal to the second preceding cumulative contribution, then further computations with respect to the chosen s (i.e., available RBs) are truncated. The system then picks the next available RBs (i.e., next s), and use possible allocated RBs (i.e., $x_k$: 0, 2, 4, 6, 100) to compute cumulative contributions to the total throughput; while checking the truncation conditions. This recursive process continues for all available RBs (i.e., s), for all possible allocated RBs (i.e., $x_k$) and for all scheduled users within a Cell coverage area.

Whereas the optimal decision on maximum cumulative contribution to the total throughput is taken at $x_k = 28$ in **Figure 2**, and truncation is effected at $x_k = 32$; same decision is taken in **Figure 3** after computing cumulative contributions for all $x_k$, (i.e., $x_k = 0, 2, ..., 100$). Similarly in **Figure 4**, the dynamic programming procedure is truncated after computing cumulative throughput (or contribution) for $x_k = 20$, since maximum cumulative contribution is obtained at $x_k = 16$. However, for the corresponding dynamic programming based graphs presented in **Figure 5**, the computations are carried out for all $x_k$ (i.e., $x_k = 0, 2, ..., 100$). In **Figure 6** on the one hand, the truncated dynamic programming based graph is truncated at $x_k = 4$, as the instantaneous maximum cumulative contribution is obtained at $x_k = 0$. On the other hand, for the corresponding dynamic programming procedure in **Figure 7**, the computations are carried out for all $x_k$ (i.e., $x_k = 0, 2, ..., 100$).

To further confirm the effectiveness and efficiency of the proposed truncated dynamic programming technique, we compare its performance with the Fair Allocation technique. For the Fair allocation technique, 6 RBs are allocated to users with good signal strength and 4 RBs are allocated to

**Table 3.** Option 1.

| Truncated Dynamic Programming | | Fair Allocation | |
|---|---|---|---|
| TTI | Cumulative Throughput | TTI | Cumulative Throughput |
| 1 | 33568 | 1 | 23632 |
| 2 | 55168 | 2 | 42464 |
| 3 | 73424 | 3 | 56552 |
| 4 | 81824 | 4 | 68544 |
| 5 | 90224 | 5 | 78000 |
| 6 | 98624 | 6 | 84912 |
| 7 | 107024 | 7 | 90544 |
| 8 | 115328 | 8 | 94496 |
| 9 | 120000 | 9 | 97520 |
| | | 10 | 100344 |
| | | 11 | 102864 |
| | | 12 | 105384 |
| | | 13 | 107856 |
| | | 14 | 109872 |
| | | 15 | 111888 |
| | | 16 | 113904 |
| | | 17 | 115688 |
| | | 18 | 116968 |
| | | 19 | 117976 |
| | | 20 | 118984 |
| | | 21 | 119744 |
| | | 22 | 120000 |

users with poor signal strength. Relevant data on the comparative analysis are presented in Tables 3-5. The graphs in **Figure 8** for Options 1, 2, and 3, show that the Truncated Dynamic Programming technique performs better than the Fair Allocation technique.

**Table 4.** Option 2.

| Truncated Dynamic Programming | | Fair Allocation | |
|---|---|---|---|
| TTI | Cumulative Throughput | TTI | Cumulative Throughput |
| 1 | 28464 | 1 | 22352 |
| 2 | 50064 | 2 | 41792 |
| 3 | 71664 | 3 | 58856 |
| 4 | 91728 | 4 | 71832 |
| 5 | 100128 | 5 | 82384 |
| 6 | 108528 | 6 | 91264 |
| 7 | 116928 | 7 | 98480 |
| 8 | 120000 | 8 | 104016 |
| | | 9 | 107456 |
| | | 10 | 109272 |
| | | 11 | 110784 |
| | | 12 | 112296 |
| | | 13 | 113760 |
| | | 14 | 114768 |
| | | 15 | 115776 |
| | | 16 | 116784 |
| | | 17 | 117792 |
| | | 18 | 118400 |
| | | 19 | 118904 |
| | | 20 | 119408 |
| | | 21 | 119912 |
| | | 22 | 120000 |

**Table 5.** Option 3.

| Truncated Dynamic Programming | | Fair Allocation | |
|---|---|---|---|
| TTI | Cumulative Throughput | TTI | Cumulative Throughput |
| 1 | 39600 | 1 | 27992 |
| 2 | 76016 | 2 | 54024 |
| 3 | 84416 | 3 | 74176 |
| 4 | 92816 | 4 | 87616 |
| 5 | 101216 | 5 | 93104 |
| 6 | 109616 | 6 | 96280 |
| 7 | 118016 | 7 | 99304 |
| 8 | 120000 | 8 | 102328 |
| | | 9 | 105120 |
| | | 10 | 107136 |
| | | 11 | 109152 |
| | | 12 | 111168 |
| | | 13 | 113184 |
| | | 14 | 114432 |
| | | 15 | 115440 |
| | | 16 | 116448 |
| | | 17 | 117456 |
| | | 18 | 118232 |
| | | 19 | 118736 |
| | | 20 | 119240 |
| | | 21 | 119744 |
| | | 22 | 120000 |

## 5. CONCLUSION

In this paper, we proposed a truncated dynamic programming under uncertainty based technique for efficient and optimal allocation of RBs in a multi-user Cellular LTE downlink. We gave the motivation for truncating the dynamic programming procedure, and also provided the effect of truncation on the recursive procedure. The simulation analysis was done under three different Options. In view of space constraint in this paper, we could not present results for all the twenty users (i.e. n = 20) considered, but we presented results for only the 18th user (i.e. j = 18). We could have taken any other user for our analysis. Furthermore, we compared the performance of the proposed truncated dynamic programming under uncertainty with the earlier proposed deterministic dynamic programming based technique. We also compared performance of the proposed technique with the Fair allocation technique. In terms of throughput, we found that the truncated dynamic programming based technique performed better than the Fair allocation methodology. Moreover, the Truncated Dynamic Programming and the earlier proposed Dynamic Programming based techniques achieved same throughput, but the Truncated Dynamic Programming based technique was more efficient as the computational time involved was significantly lower. Thus, it is shown by the results obtained that the proposed technique is more efficient in allocating radio resource and has better performance than both the Fair Allocation and deterministic dynamic programming based techniques for scheduling in Cellular LTE downlink. Besides, the proposed technique has the added advantage of managing uncertainty in selecting applicable modulation scheme (i.e., QPSK, 16QAM or 64QAM) as a result of users' mobility within a Cell coverage area.

## REFERENCES

1. 3GPP TS 36.211. "Evolved Universal Terrestrial Radio Access (EUTRA); Physical Channel and Modulation (Release 8)," Technical Report, 3GPP-TSG R1, September 2007.
2. O. Iosif and I. Banica, "On the Analysis of Packet Scheduling in Downlink 3GPP LTE System," Proceedings of the 4th International Conference on Communication Theory, Reliability and Quality of Service, Budapest, 17-22 April 2011, pp. 99-102.

3. H. Holma and A. Toskala, "LTE for UMTS: OFDMA and SC-FDMA Based Radio Access," John Wiley & Sons, New York, 2009.
4. J. Huang, V. G. Subramanian, R. Agrawal and R. A. Berry, "Downlink Scheduling and Resource Allocation for OFDM Systems," IEEE Transactions on Wireless Communications, Vol. 8, No.1, 2009, pp. 288-296.
5. M. S. Kaiser and K. M. Ahmed, "Radio Resource Allocation for Heterogeneous Services in Relay Enhanced OFDMA Systems," Journal of Communications, Vol. 5, No. 6, 2010, pp. 447-454.
6. A. Wang, Y.-Y. Qiu, L. Lin and S. Li, "An Adaptive Sub-Carrier and Power Allocation Algorithm with QoS Guarantee for OFDMA System," 10th IEEE Conference on High Performance Computing and Communications, Dalian, 25-27 September 2008, pp. 492-497.
7. G. H. Kumar, P. H. Krishnarao and K. S. R. Krishna, "Two-Stage Algorithm for Sub-Carrier, Bit and Power Allocation in OFDMA Systems," International Journal of Computer Science and Communications, Vol. 1, No. 2, 2010, pp. 51-54.
8. G. Li and H. Liu, "Downlink Radio Resource Allocation for Multi-Cell OFDMA System," IEEE Transactions on Wireless Communications, Vol. 5, No. 12, 2006, pp. 3451- 3459.doi:10.1109/TWC.2006.256968
9. I. Koutsopoulos and L. Tassiulas, "Cross-Layer Adaptive Techniques for Throughput Enhancement in Wireless OFDM-Based Networks," IEEE/ACM Transactions on Networking, Vol. 14, No. 5, 2006, pp. 1056-1066.
10. T. Thonabalasinghan, S. Hanley, L. L. H. Andrew and J. Papandripoulos, "Joint Allocation of Sub-Carriers and Transmit Powers in a Multi-User OFDM Cellular Network," IEEE International Conference on Communications, Istanbul, 11-15 June 2006, pp. 269-274.
11. C. Koutsimanis and G. Fodor, "A Dynamic Resource Allocation Scheme for GBR Services in OFDMA Networks," Proceedings of the IEEE Conference of Communications, Beijing, 19-23 May 2008, pp. 2524-2530.
12. I. C. Wong and B. L. Evans, "Optimal Resource Allocation in OFDMA Systems with Imperfect Channel Knowledge," IEEE Transactions on Communications, Vol. 57, No. 1, 2009, pp. 232-241.

13. P. Hosein, "Coordinated Radio Resource Management for the LTE Downlink: The Two-Sector Case," IEEE International Conference on Communications, Cape Town, 23-27 May 2010, pp. 1-5.
14. N. Zhou, X. Zhu and Y. Huang, "Genetic Algorithm Based Cross-Layer Resource Allocation for Wireless OFDM Networks with Heterogeneous Traffic," 17th European Signal Processing Conference (EUSIPCO), Glasgow, 24- 28 August 2009, pp. 1656-1659.
15. M. Ergen, S. Coleri and P. Varaiya, "QoS Aware Adaptive Resource Allocation Techniques for Fair Scheduling in OFDMA Based Broadband Wireless Access Systems," IEEE Transactions on Broadcasting, Vol. 49, No. 4, 2003, pp. 362-370.
16. A. M. Ajofoyinbo and K. Orolu, "A Dynamic Programming Based Technique for Optimal Allocation of Radio Resource in Multi-User Cellular Long-Term Evolution (LTE) Downlink," 10th WSEAS International Conference on Telecommunications and Informatics (TELE-INFO'11), Lanzarote, Spain, 27-29 May 2011, pp. 107-113.
17. F. S. Hillier and G. J. Lieberman, "Introduction to Mathematical Programming," 2nd Edition, McGraw-Hill Inc., New York, 1995.

# CHAPTER 8

## VARIATIONAL BAYESIAN MIXED-EFFECTS INFERENCE FOR CLASSIFICATION STUDIES

Kay H. Brodersen[a, b, c,], Jean Daunizeau[c, d], Christoph Mathys[a, c], Justin R. Chumbley[c], Joachim M. Buhmann[b], Klaas E. Stephan[a, c, e]

[a] Translational Neuromodeling Unit (TNU), Institute for Biomedical Engineering, University of Zurich & ETH Zurich, Switzerland

[b] Machine Learning Laboratory, Department of Computer Science, ETH Zurich, Switzerland

[c] Laboratory for Social and Neural Systems Research (SNS), Department of Economics, University of Zurich, Switzerland

[d] Institut du Cerveau et de la Moelle Épinière (ICM), Hôpital Pitié Salpêtrière, Paris, France

[e] Wellcome Trust Centre for Neuroimaging, University College London, UK

## ABSTRACT

Multivariate classification algorithms are powerful tools for predicting cognitive or pathophysiological states from neuroimaging data. Assessing the utility of a classifier in application domains such as cognitive neuroscience, brain–computer interfaces, or clinical diagnostics necessitates inference on classification performance at more than one level, i.e., both in individual subjects and in the population from which these subjects were sampled. Such inference requires models that explicitly account for both fixed-effects (within-subjects) and random-effects (between-subjects) variance components. While models of this sort are standard in mass-univariate analyses of fMRI data, they have not yet received much attention in multivariate classification studies of neuroimaging data, presumably because of the high computational costs they entail. This paper extends a recently developed hierarchical model for mixed-effects inference in multivariate classification studies and introduces an efficient variational Bayes approach to inference. Using both synthetic and empirical fMRI data, we show that this approach is equally simple to use as, yet more powerful than, a conventional $t$-test on subject-specific sample accuracies, and computationally much more efficient than previous sampling algorithms and permutation tests. Our approach is independent of the type of underlying classifier and thus widely applicable. The present framework may help establish mixed-effects inference as a future standard for classification group analyses.

## KEYWORDS

Variational Bayes; Fixed effects; Random effects; Normal-binomial; Balanced accuracy; Bayesian inference; Group studies

## 1. INTRODUCTION

Multivariate classification algorithms have emerged from the field of machine learning as powerful tools for predicting cognitive or pathophysiological states from neuroimaging data (Haynes and Rees, 2006). Classifiers are based on decoding models that differ in two ways from conventional mass-univariate encoding analyses based on the general linear model (GLM; Friston et al., 1995). First, multivariate approaches explicitly account for dependencies among voxels. Second, they reverse the direction of inference, predicting a contextual variable from brain activity (decoding) rather than the other way around (encoding). There are three related areas of application in which these two characteristics have sparked most interest.

In cognitive neuroscience, and in particular neuroimaging, classifiers have been employed to decode subject-specific cognitive or perceptual states from multivariate measures of brain activity, such as those obtained by fMRI (Brodersen et al., 2012b, Cox and Savoy, 2003, Haynes and Rees, 2006, Norman et al., 2006 and Tong and Pratte, 2012). A second area is the design of brain–machine interfaces which aim at decoding subjective cognitive states (e.g., intentions or decisions) from trial-wise measurements of neuronal activity in individual subjects (Blankertz et al., 2011 and Sitaram et al., 2008). A third important domain concerns clinical applications that explore the utility of multivariate decoding approaches for diagnostic purposes (Davatzikos et al., 2008, Klöppel et al., 2008, Klöppel et al., 2012 and Marquand et al., 2010). Recently, decoding models have also been integrated with biophysical models of brain function, such as dynamic causal models (Friston et al., 2003), to afford mechanistically interpretable classifications (Brodersen et al., 2011a and Brodersen et al., 2011b).

Many applications of multivariate classification operate on data with a two-level hierarchical structure. Consider, for example, a study in which a classification algorithm is used to decode from fMRI data whether a subject chose option A or B on each of $n$ experimental repetitions or trials. This analysis gives rise to $n$ estimated labels (representing which choice the classifier predicted on each trial) and $n$ true labels (indicating which option was truly chosen). Comparing predicted to true labels yields a sequence of classification *outcomes* (indicating for each trial whether the prediction was correct or incorrect). Repeating this analysis for each member of a group of $m$ subjects yields the typical two-level structure ($m$ subjects times $n$ trials each) that is illustrated in Fig. 1; for a concrete example see Figs. 7a,e. A two-level structure underlies virtually all trial-by-trial decoding studies (see, among many others, Brodersen et al., 2012b, Chadwick et al., 2010, Harrison and Tong, 2009, Johnson et al., 2009 and Krajbich et al., 2009). The same two-level structure often applies to subject-by-subject classification studies (e.g., decoding a diagnostic state or predicting a clinical outcome), especially when subjects are partitioned into groups that are analyzed separately.

A hierarchical (or multilevel) design of this sort gives rise to the questions of what we can infer about the accuracy of the classifier in individual subjects, and what about the accuracy in the population from which the subjects were sampled. Any approach to answering these questions must provide a means of (i) *estimation* (e.g., of the accuracy itself as well as an appropriate interval that describes our uncertainty about the accuracy); and (ii) *testing* (e.g., whether the accuracy is significantly above chance). This paper is concerned with such subject-level and group-level inferences on classification accuracy for multilevel data.

The statistical evaluation of classification performance in non-hierarchical (e.g., single-subject) applications of classification has been discussed extensively in the literature (Brodersen et al., 2010a, Langford,

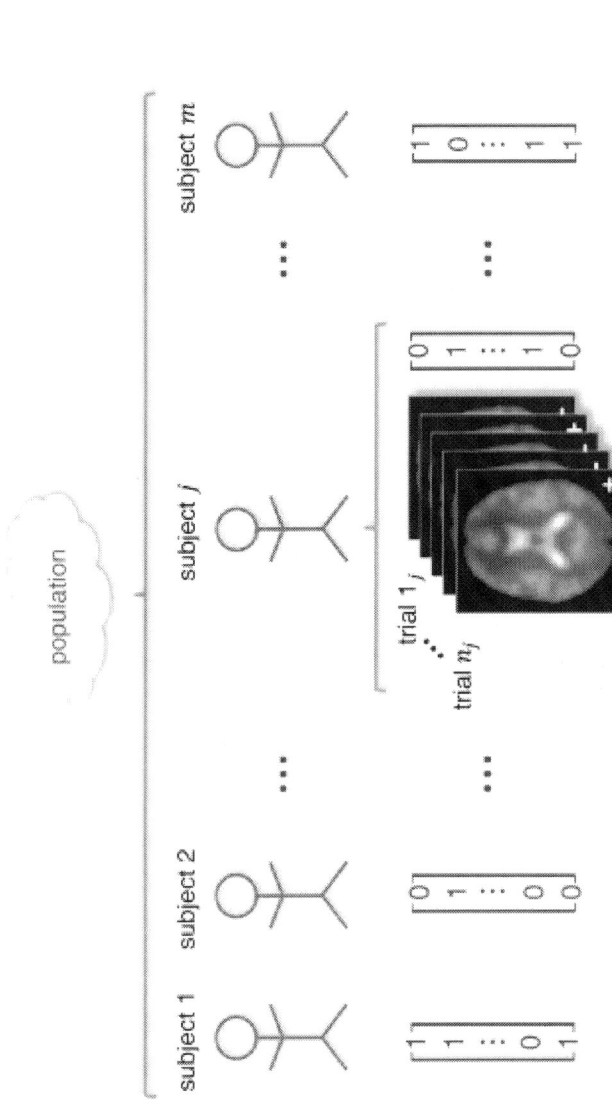

**Figure 1.** Overview of the outcomes generated by a classification group study. In a trial-by-trial classification analysis, a classifier is trained and tested, separately for each subject, to predict a binary label (+ or −) from trial-wise correlates of brain activity. This constitutes a hierarchical design. The first level concerns trial-wise classification outcomes (where 1 and 0 represent correctly and incorrectly classified trials) that are drawn from latent subject-specific classification accuracies. The second level concerns subject-specific accuracies themselves, which are drawn from a population distribution. When evaluating the performance of a classification algorithm, we are interested in inference on subject-specific accuracies and on the population accuracy itself.

2005, Lemm et al., 2011, Pereira and Botvinick, 2011 and Pereira et al., 2009). By contrast, relatively little attention has thus far been devoted to evaluating classification algorithms in hierarchical (i.e., group) settings (Goldstein, 2010 and Olivetti et al., 2012). This is unfortunate since the field would benefit from a broadly accepted standard.

Such a standard approach to evaluating classification performance in a hierarchical setting should account for two independent sources of variability: *fixed-effects* (i.e., within-subjects) variance that results from uncertainty about the true classification accuracy in any given subject; and *random-effects* variance (i.e., between-subjects variability) that reflects the distribution of true accuracies in the population from which subjects were sampled. This distinction is crucial because classification outcomes obtained in different subjects cannot be treated as samples from the same distribution; in a hierarchical setting, each subject itself has been sampled from a population with an unknown intrinsic heterogeneity ( Beckmann et al., 2003 and Friston et al., 2005). Models that explicitly separate both sources of uncertainty are known as *mixed-effects* models. They are the objects of interest in this paper.

Contemporary approaches to performance evaluation in classification group studies fall into several groups.[1] One approach rests on the *pooled sample accuracy*, i.e., the number of correctly predicted trials, summed across all subjects, divided by the overall number of trials. The statistical significance of the pooled sample accuracy can be assessed using a simple classical binomial test (assuming the standard case of binary classification) that is based on the likelihood of obtaining the observed number of correct trials (or more) by chance ( Langford, 2005). A less frequent variant of this analysis uses the *average sample accuracy* instead of the pooled sample accuracy ( Clithero et al., 2011).

A second approach, more commonly used, is to consider *subject-specific sample accuracies* and estimate their distribution in the population. This method typically (explicitly or implicitly) uses a classical one-tailed $t$-test across subjects to assess whether the population mean accuracy is greater than what would be expected by chance (e.g., Harrison and Tong, 2009, Knops et al., 2009, Krajbich et al., 2009 and Schurger et al., 2010).

In the case of single-subject studies, the first method (i.e., a binomial test on the pooled sample accuracy) is an appropriate approach. However, there are three reasons why neither method is optimal for group studies. Firstly, both of the above methods neglect the hierarchical nature of the experiment. The first method (based on the pooled sample accuracy) represents a fixed-effects approach and disregards variability across subjects. This leads to overly optimistic inferences and provides results that are only representative for the specific sample of subjects studied, not for the population they were drawn from. The second method ($t$-test on sample accuracies) does consider random effects; but it neither explicitly models the uncertainty associated with subject-specific accuracies, nor does it account for violations of homoscedasticity (i.e., the differences in variance of the data between subjects).

The second limitation of the above methods is rooted in their distributional assumptions. In the standard case of binary classification, it is reasonable to assume individual classification outcomes to follow binomial distributions (justifying the binomial test in single-subject studies). However, it is not well founded to assume that sample accuracies follow a Gaussian distribution (which, in this particular case, is the implicit assumption of a classical $t$-test on sample accuracies). This is because a Gaussian has infinite support, which means it inevitably places probability mass on values below 0% and above 100% (for an alternative, see Dixon, 2008).

A third problem, albeit not an intrinsic characteristic of the above methods, is their typical focus on classification accuracy, which is known to be a poor indicator of performance when classes are not perfectly balanced. Specifically, a classifier trained on an imbalanced dataset may acquire a bias in favor of the majority class, resulting in an overoptimistic accuracy. This motivates the use of an alternative performance measure, the *balanced accuracy*, which removes this bias from performance evaluation.

We recently proposed a solution to the three above limitations using Bayesian hierarchical models for mixed-effects inference on classification performance. In particular, we introduced the *beta-binomial* model and the *normal-binomial* model for inferring on both accuracies and balanced accuracies ( Brodersen et al., 2012a). Both models use a fully Bayesian framework for mixed-effects inference, are based on natural distributional assumptions, and enable more accurate inferences than the two conventional approaches described earlier. The models are independent of the type of underlying classifier, which makes them widely applicable.

The practical utility of our models, however, has been limited by the high computational complexity of the underlying Markov chain Monte Carlo (MCMC) sampling algorithms required for model inversion (i.e., the process of passing from a prior to a posterior distribution over model parameters, given the data). MCMC is asymptotically exact; but it is also exceedingly slow, especially when performing inference in a voxel-by-voxel fashion, as is common, for example, in 'searchlight' approaches (Kriegeskorte et al., 2006 and Nandy and Cordes, 2003).

In this paper, we present a variational Bayes (VB) algorithm to overcome this critical limitation.[2] Our approach has three main features. First, we present a mixed-effects model that explicitly respects the hierarchical structure of the data. Second, the model can be equally used for inference on the accuracy and the balanced accuracy. Third, our novel variational inference scheme dramatically reduces the computational complexity (i.e., runtime) compared to our previous sampling approach based on MCMC.

The paper is organized as follows. In the Theory section, we present variations of our recently developed normal-binomial model for mixed-effects inference (Brodersen et al., 2012a). These are the *univariate normal-binomial* model (for inference on the *accuracy*) and the *twofold normal-binomial* model (for inference on the *balanced accuracy*).[3] We then describe a novel VB algorithm for model inversion and compare it to an MCMC sampler. In the Applications section, we provide a set of illustrative results on both synthetic data and empirical fMRI measurements. Finally, in the Discussion, we review the key characteristics of our approach, compare it to similar models in other analysis domains, and discuss its role in future classification studies.

## 2. THEORY

In a hierarchical setting, a classifier is typically used to predict a class label for each trial, where trials are further structured into sets, for instance because they were recorded from different subjects. The most common situation is binary classification, where class labels are taken from $\{+1, -1\}$, denoting 'positive' and 'negative' trials, respectively. Less common, but equally amenable to the approach presented in this paper, are multiclass settings in which trials fall into more than two classes (see Discussion).

The above situation raises three principal questions (cf. Brodersen et al., 2012a). First, can one obtain successful classification at the group level? This requires statistical inference on the mean classification accuracy in the population from which subjects were drawn. Second, do the subject-wise data permit classification in each individual? Considering each subject in isolation is statistically short-sighted, since subject-specific inference may benefit from simultaneous across-subject inference (Efron and Morris, 1971). Third, which of several possible classification algorithms should be chosen? This is typically answered by evaluating how well an algorithm's performance generalizes (to unseen data). In a Bayesian framework, this expected performance is given by the posterior predictive density of classification performance. The present section describes a variational Bayes (VB) approach to answering these questions (Fig. 2).

The univariate normal-binomial model for inference on the accuracy

Within each subject, classification outcomes can be summarized in terms of the number of correctly predicted trials, $k$, and the total number of trials, $n$. It is important to note that this summary is independent of the type of underlying classifier. This means that the model can be applied regardless of whether classification results were obtained using, for instance, logistic regression, nearest-neighbor classification, a support vector machine, or a Gaussian process classifier. Under the assumption that trial-specific predictions are conditionally independent, $k$ follows a binomial distribution,

$$p(k|\pi, n) = \text{Bin}(k|\pi, n) = \binom{n}{k} \pi^k (1-\pi)^{n-k} \qquad (1)$$

where $\pi$ represents the latent (unobservable) accuracy of the classifier, $0 \leq \pi \leq 1$. Thus, in a group study, where the classifier has been trained and tested separately in each subject, the available data are $kj$ and $nj$ for each subject $j = 1...m$.

One might be tempted to form group summaries $k = \sum j_{=1mkj}$ and $n = \sum j_{=1mnj}$ and proceed to inference on $\pi$. However, using such a *pooled sample accuracy* would assume zero between-subjects variability. In other words, $\pi$ would be treated as a *fixed effect* in the population. This approach would not permit inferences about the population; it would only allow for results to be reported as a case study ( Friston et al., 1999).

Alternatively, one might summarize the data from each subject in terms of a subject-specific *sample accuracy*, $kj/nj$. One could then ask, using a one-tailed $t$-test, whether sample accuracies reflect a normal distribution with a mean greater than what would be expected by chance ( Fig. 2a). This approach no longer treats accuracy as a fixed effect. However, it suffers from two other problems.

First, submitting subject-specific sample accuracies to a $t$-test assumes that accuracies, which are confined to the [0,1] interval, follow a normal distribution, which has infinite support. This may lead to non-interpretable

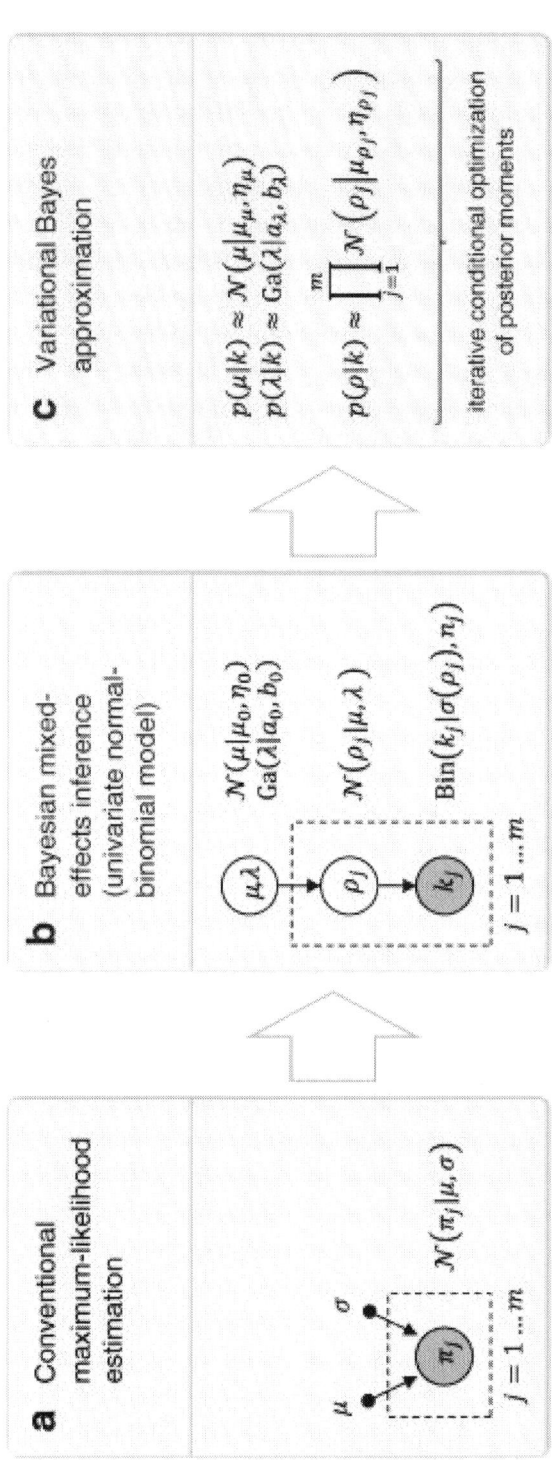

**Figure 2.** Inference on classification accuracies. (a) Conventional maximum-likelihood estimation does not explicitly model within-subjects (fixed-effects) variance components and is based on an ill-justified normality assumption. It is therefore inadequate for the statistical evaluation of classification group studies. (b) The normal-binomial model respects the hierarchical structure of the study and makes natural distributional assumptions, thus enabling mixed-effects inference, which makes it suitable for group studies. The model uses the sigmoid transform $\sigma(\rho j) := (1 + \exp(-\rho j))^{-1}$ which turns log-odds with real support $(-\infty, \infty)$ into accuracies on the $[0,1]$ interval. (b) Model inversion can be implemented efficiently using a variational Bayes approximation to the posterior densities of the model parameters (see Fig. 3 for details).

results such as confidence intervals that include accuracies above 100% (or below 0%). [4]

Second, even if one were to overcome the above problem (e.g., using a logit transform), a *t*-test on sample accuracies neither explicitly accounts for within-subjects uncertainty nor for violations of homoscedasticity. This is because it uses sample accuracies as summary statistics without carrying forward the uncertainty associated with them (Mumford and Nichols, 2009). For example, sample accuracies do not distinguish between an accuracy of 80% that was obtained as 80 correct out of 100 trials (i.e., an estimate with high confidence) and the same accuracy obtained as 8 out of 10 trials (i.e., an estimate with low confidence). Furthermore, no distinction regarding the confidence in the inference is being made between 80 correct out of 100 trials (i.e., high confidence) and 50 correct out of 100 trials (lower confidence, since the variance of a binomial distribution depends on its mean and becomes maximal at a mean of 0.5).

In order to explicitly capture both within-subjects (fixed-effects) and between-subjects (random-effects) variance components, we must instead use a hierarchical model in which separate levels account for different sources of variability (Fig. 2b). At the level of individual subjects, for each subject *j*, the number of correctly classified trials $k_j$ is modeled as

$$p(k_j|\pi_j, n_j) = \text{Bin}(k_j|\pi_j, n_j) \qquad (2)$$

where $\pi_j$ represents the latent classification accuracy in subject *j*. [5] Next, at the group level, we account for variability between subjects by modeling subject-specific accuracies as drawn from a population distribution. The *natural* parameter of the binomial density is $\ln \frac{\pi}{1-\pi}$. Thus, one possible parameterization is to assume accuracies to be logit-normally distributed and conditionally independent given the population parameters. In other words, each logit accuracy $= \sigma^{-1}(\pi_j) := \ln \frac{\pi_j}{1-\pi_j}$ is drawn from a normal distribution. The inverse-sigmoid (or logit) transform $\sigma^{-1}(\pi_j)$ turns accuracies with support on the [0,1] interval into log-odds with support on the real line $(-\infty, +\infty)$. Thus,

$$p(\rho_j|\mu, \lambda) = N(\rho_j|\mu, \lambda) = \sqrt{\frac{\lambda}{2\pi}} \exp\left(-\frac{\lambda}{2}(\rho_j - \mu)^2\right) \qquad (3)$$

where $\mu$ and $\lambda$ represent the population mean and the population precision (i.e., inverse variance), respectively.

Since neuroimaging studies are typically confined to relatively small sample sizes, an adequate expression of our prior ignorance about the population parameters is critical (cf. Woolrich et al., 2004). We use a diffuse prior on $\mu$ and $\lambda$ such that the posterior will be dominated by the data (for a validation of this prior, see Applications). A straightforward parameterization is to use independent conjugate densities:

$$p(\mu|\mu_0,\eta_0)=N(\mu|\mu_0,\eta_0). \tag{4}$$

$$p(\lambda|a_0,b_0)=Ga(\lambda|a_0,b_0). \tag{5}$$

In the above densities, $\mu_0$ and $\eta_0$ encode the prior mean and precision of the population mean, and $a_0$ and $b_0$ represent the shape and scale parameter,[6] respectively, that specify the prior distribution of the population precision (for an alternative, see Leonard, 1972). In summary, the univariate normal-binomial model uses a binomial distribution at the level of individual subjects and a logit-normal distribution at the group level (Fig. 2b).

In principle, inverting the above model immediately yields the desired posterior density over parameters,

$$p(\mu,\lambda,\rho|k) = \frac{\prod_{j=1}^{m}\left[\text{Bin}\left(k_j|\sigma(\rho_j)\right)N\left(\rho_j\middle|\mu,\lambda\right)\right]N(\mu|\mu_0,\eta_0)Ga(\lambda|a_0,b_0)}{p(k)} \tag{6}$$

In practice, however, integrating the expression in the denominator of the above expression, which provides the normalization constant for the posterior density, is prohibitively difficult. We previously described a stochastic approximation based on MCMC algorithms; however, the practical use of these algorithms was limited by their considerable computational complexity (Brodersen et al., 2012a). Here, we propose to invert the above model using a deterministic VB approximation (Fig. 2c). This approximation is no longer asymptotically exact, but it conveys considerable compu-

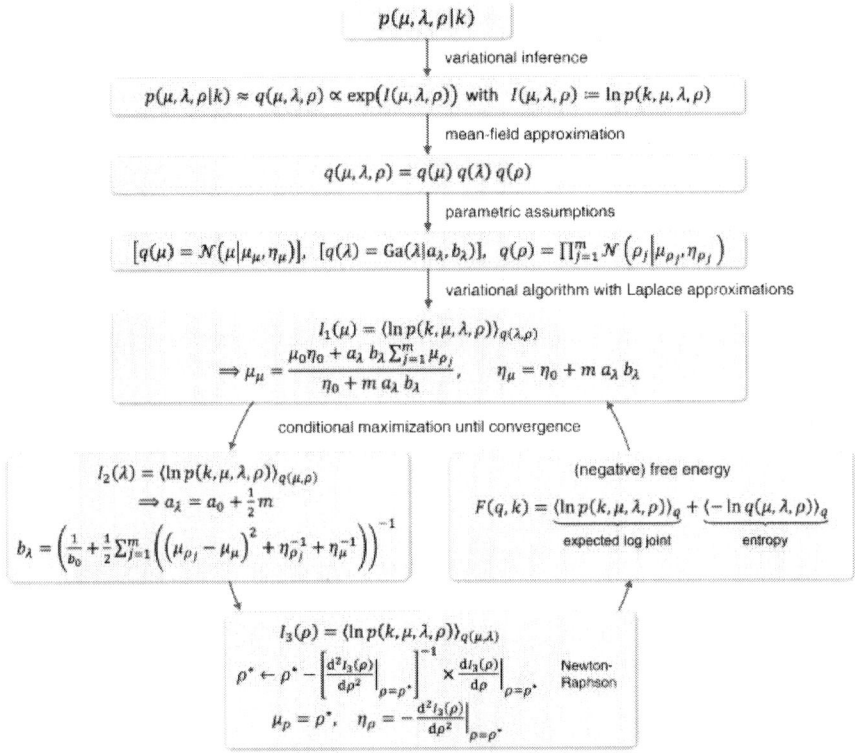

**Figure 3.** Variational inversion of the univariate normal-binomial model. This schematic summarizes the individual steps involved in the variational approach to the inversion of the univariate normal-binomial model, as described in the main text (see Theory).

tational advantages. The remainder of this section describes its derivation (see Fig. 3 for a summary).

## 3. VARIATIONAL INFERENCE

The difficult problem of finding the exact posterior $p(\mu,\lambda,\rho|k)$ can be transformed into the easier problem of finding an approximate parametric posterior $q(\mu,\lambda,\rho|\delta)$ with moments (i.e., parameters) $\delta$. (We will omit $\delta$ to simplify the notation.) Inference then reduces to finding a density $q$ that minimizes a measure of dissimilarity between $q$ and $p$. This can be achieved by maximizing the so-called negative free energy $F$ of the model, a lower-bound approximation to the log model evidence, with respect to (the moments of) $q$. For details, see MacKay (1995), Attias (2000), Ghahramani and Beal (2001), Bishop et al. (2002), and Fox and Roberts (2012). Maximizing the negative free energy minimizes the Kullback–Leibler (KL) divergence between the approximate and the true posterior, $q$ and $p$:

$$\mathrm{KL}[q\|p] := \iiint q(\mu,\lambda,\rho) \ln \frac{q(\mu,\lambda,\rho)}{p(\mu,\lambda,\rho|k)} d\mu d\lambda d\rho \tag{7}$$

$$= \iiint q(\mu,\lambda,\rho) \ln \frac{q(\mu,\lambda,\rho)}{p(k,\mu,\lambda,\rho)} d\mu d\lambda d\rho + \ln p(k) \tag{8}$$

$$\Leftrightarrow \ln p(k) = \mathrm{KL}[q\|p] + \underbrace{\left\langle \ln \frac{p(k,\mu,\lambda,\rho)}{q(\mu,\lambda,\rho)} \right\rangle_{q(\mu,\lambda,\rho)}}_{=:F(q,k)} \tag{9}$$

This means that the log-model evidence $\ln p(k)$ can be expressed as the sum of (i) the KL-divergence between the approximate and the true posterior and (ii) the negative free energy $F(q,k)$. Because the KL-divergence cannot be negative, maximizing the negative free energy with respect to $q$ minimizes the KL-divergence and thus results in an approximate posterior that is maximally similar to the true posterior. At the same time, maximizing the negative free energy provides a lower-bound approximation to the log-model evidence, which permits Bayesian model comparison (Bishop, 2007 and Penny et al., 2004). In summary, maximizing the negative free energy $F(q,k)$ in Eq. (9) enables both inference on the posterior density over parameters and model comparison. In this paper, we are primarily interested in the posterior density.

In trying to maximize $F(q,k)$, variational calculus tells us that

$$\frac{\partial F(q,k)}{\partial q} = 0 \Rightarrow q(\mu,\lambda,\rho) \propto \exp[\underbrace{\ln p(k,\mu,\lambda,\rho)}_{\text{negative variational energy}}] \tag{10}$$

This means that the approximate posterior which maximizes the negative free energy is equal to the true posterior and thus proportional to the joint density over data and parameters[7] (with the normalization constant being given by the model evidence). In other words, the VB approach is complete in the sense that, in the absence of any other approximations, optimizing $F$ with respect to $q$ yields the exact posterior density and model evidence.

## 4. MEAN-FIELD APPROXIMATION

To make the optimization on the l.h.s. in Eq. (10) tractable, we assume that the joint posterior over all model parameters factorizes into specific parts. Using one density for each variable,

$$q(\mu, \lambda, \rho) = q(\mu)q(\lambda)q(\rho) \tag{11}$$

the mean-field assumption turns the problem of maximizing $F(q,k)$ into the problem of deriving three expectations:

$$I_1(\mu) = \langle \ln p(k, \mu, \lambda, \rho) \rangle_{q(\lambda,\rho)} \tag{12}$$

$$I_2(\lambda) = \langle \ln p(k, \mu, \lambda, \rho) \rangle_{q(\mu,\rho)} \tag{13}$$

$$I_3(\rho) = \langle \ln p(k, \mu, \lambda, \rho) \rangle_{q(\mu,\lambda)}. \tag{14}$$

This transformation has several advantages over working with Eq. (10) directly: it makes it more likely that we can find the exact distributional form of a marginal approximate posterior (as will be the case for $\mu$ and $\lambda$); it may make the Laplace assumption more appropriate in those cases where we cannot identify a fixed form (as will be the case for $\rho$); and it often provides us with interpretable update equations (as will be the case, in particular, for $\mu$ and $\lambda$).

## 5. PARAMETRIC ASSUMPTIONS

Due to the structure of the model, the posteriors on the population parameters $\mu$ and $\lambda$ are conditionally independent given the data. In addition, owing to the conjugacy of their priors, the posteriors on $\mu$ and $\lambda$ follow the same distributions and do not require any additional parametric assumptions:

$$q(\mu) = N\left(\mu | \mu_\mu, \eta_\mu\right) \tag{15}$$

$$q(\lambda) = Ga(\lambda | a_\lambda, b_\lambda). \tag{16}$$

Subject-specific (logit) accuracies $q \equiv (\rho_1,\ldots,\rho m)$ are also conditionally independent given the data. This is a consequence of the fact that the posterior for each subject only depends on its Markov blanket, i.e., the subject's data and the population parameters (but not the other subject's logit accuracies). This can be seen from the fact that

$$q(\mu, \lambda, \rho) = q(\mu)\,q(\lambda)\,q(\rho) \tag{17}$$

$$= q(\mu)q(\lambda)\prod_{j=1}^{m} q\left(\rho_j\right). \tag{18}$$

However, we do require a distributional assumption for the above subject-specific posteriors to make model inversion feasible. Here, we assume posterior subject-specific (logit) accuracies to be normally distributed:

$$q(\rho) = \prod_{j=1}^{m} N\left(\rho_j | \mu_{\mu_j}, \eta_{\rho_j}\right). \tag{19}$$

The conditional independence in Eq. (19) differs in a subtle but important way from the assumption of *unconditional* independence that is implicit in random-effects analyses on the basis of a *t*-test on subject-specific sample accuracies (see Introduction). In the case of such *t*-tests, estimation in each subject only ever uses data from that same subject. By contrast, the subject-specific posteriors in Eq. (20) borrow strength from *all* observations. This can be seen from the fact that the subject-specific posteriors $q(\rho)$ are computed with respect to the population posteriors $q(\mu)$ and $q(\lambda)$ which are themselves informed by observations from the entire group (see Eqs. (12), (13) and (14)).

## 6. DERIVATION OF VARIATIONAL DENSITIES

For each mean-field part in Eq. (11), the variational density $q(\cdot)$ can be obtained by evaluating the variational energy $I(\cdot)$, as described next. The first variational energy concerns the posterior density over the population mean $\mu$. It is given by

$$I_1(\mu) = \langle \ln p(k, \mu, \lambda, \rho) \rangle_{q(\lambda,\rho)} \tag{20}$$

$$= \langle \ln p(k|\rho) \rangle_{q(\lambda,\rho)} + \langle \ln p(\rho|\mu, \lambda) \rangle_{q(\lambda,\rho)} + \langle \ln p(\mu, \lambda) \rangle_{q(\lambda,\rho)} \tag{21}$$

$$= \sum_{j=1}^{m} \langle \ln N(\rho_j|\mu, \lambda) \rangle_{q(\lambda,\rho)} + \langle \ln(N(\mu|\mu_0, \eta_0) Ga(\lambda|a_0, b_0)) \rangle_{q(\lambda,\rho)} + c \tag{22}$$

$$= \sum_{j=1}^{m} \left\langle \tfrac{1}{2} \ln \lambda - \tfrac{1}{2} \ln 2\pi - \tfrac{\lambda}{2}(\rho_j - \mu)^2 \right\rangle_{q(\lambda,\rho)} + \left\langle \tfrac{1}{2} \ln \eta_0 - \tfrac{\eta_0}{2}(\mu - \mu_0)^2 \right\rangle_{q(\lambda,\rho)} + c \tag{23}$$

$$= \sum_{j=1}^{m} -\tfrac{1}{2} \langle \lambda \rho_j^2 - 2\lambda \rho_j \mu + \lambda \mu^2 \rangle_{q(\lambda,\rho)} - \tfrac{\eta_0}{2}(\mu - \mu_0)^2 + c \tag{24}$$

$$= -\tfrac{1}{2} \sum_{j=1}^{m} \left[ -2\mu \mu_{\rho_j} + \mu^2 \right] a_\lambda b_\lambda + \mu \eta_0 \left( \mu_0 - \tfrac{1}{2}\mu \right) + c \tag{25}$$

$$= \mu a_\lambda b_\lambda \left( -\frac{1}{2} m\mu + \sum_{j=1}^{m} \mu_{p_j} \right) + \mu \eta_0 \left( \mu_0 - \frac{1}{2}\mu \right) + c \tag{26}$$

where the symbol $c$ is used for any expression that is constant with respect to $\mu$.

In principle, we could proceed by optimizing the sufficient statistics of the approximate posterior. Instead, we only optimize the mean and equate the variance to the *observed information*, i.e., the negative curvature at the mode. This procedure is known as the *Laplace approximation* (or normal approximation) and implies that the negative free energy is a function simply of the posterior means (as opposed to a function of the posterior means and covariances). It is a local, rather than a global, optimization solution.

Conveniently, the Laplace approximation is typically more accurate for the conditional posterior (of one parameter given the others) than for the full posterior (of all parameters). In addition, it is computationally efficient (see Discussion) and often gives rise to interpretable update equations (see below).

Setting the first derivative to zero yields an analytical expression for the maximum,

$$\frac{dI_1(\mu)}{d\mu} = -\mu(\eta_0 + m a_\lambda b_\lambda) + \mu_0 \eta_0 + a_\lambda b_\lambda \sum_{j=1}^{m} \mu_{p_j} = 0 \tag{27}$$

$$\Rightarrow \mu^* = \frac{\mu_0 \eta_0 + a_\lambda b_\lambda \sum_{j=1}^{m} \mu_{p_j}}{\eta_0 + m a_\lambda b_\lambda}. \tag{28}$$

Having found the mode of the approximate posterior, we can use a second-order Taylor expansion to obtain closed-form approximations for its moments:

$$\mu_\mu = \mu^* \quad \text{and} \tag{29}$$

$$\eta_\mu = -\left. \frac{dI_1^2(\mu)}{d\mu^2} \right|_{\mu=\mu^*} = \eta_0 + m a_\lambda b_\lambda. \tag{30}$$

Thus, the posterior density of the population mean logit accuracy under our mean-field and Gaussian approximations is $N(\mu|\mu\mu,\eta\mu)$.

The use of a Laplace approximation, as we do here, often leads to interpretable update equations. In Eq. (30), for example, we can see that the posterior precision of the population mean ($\eta\mu$) is simply the sum of the prior precision ($\eta_0$) and the mean of the posterior population precision ($a\lambda b\lambda$), correctly weighted by the number of subjects $m$.

Based on the above approximation for the posterior logit accuracy, we can see that the posterior mean accuracy itself, $\xi := \sigma(\mu)$, is logit-normally distributed and can be expressed in closed form,

$$\text{logit} N\left(\xi | \mu_\mu, \eta_\mu\right) = \frac{1}{\xi(1-\xi)}\sqrt{\frac{\eta_\mu}{2\pi}}\exp\left(-\frac{\eta_\mu}{2}\left(\sigma^{-1}(\xi)-\mu_\mu\right)^2\right) \quad (31)$$

where $\mu\mu$ and $\eta\mu$ represent the posterior mean and precision, respectively, of the population mean logit accuracy.

The second variational energy concerns the population precision $\lambda$ and is given by

$$I_2(\lambda) = \langle \ln p(k,\mu,\lambda,\rho)\rangle_{q(\mu,\rho)} \quad (32)$$

$$= \frac{m}{2}\ln\lambda - \frac{\lambda}{2}\sum_{j=1}^m\left(\left(\mu_{\rho_j}-\mu_\mu\right)^2 + \eta_{\rho_j}^{-1} + \eta_\mu^{-1}\right) + (a_0-1)\ln\lambda - \frac{\lambda}{b_0} + c \quad (33)$$

where $c$ represents a term that is constant with respect to $\lambda$. The above expression already has the form of a log-Gamma distribution with parameters

$$a_\lambda = a_0 + \frac{1}{2}m \quad \text{and} \quad (34)$$

$$b_\lambda = \left(\frac{1}{b_0} + \frac{1}{2}\sum_{j=1}^m\left(\left(\mu_{\rho_j}-\mu_\mu\right)^2 + \eta_{\rho_j}^{-1} + \eta_\mu^{-1}\right)\right)^{-1}. \quad (35)$$

From this we can see that the shape parameter $a\lambda$ is a weighted sum of prior shape $a_0$ and data $m$. When viewing the second parameter as a 'rate' coefficient $b\lambda^{-1}$ (as opposed to a shape coefficient $b\lambda$), it becomes clear that the posterior rate really is a weighted sum of: the prior rate ($b_0^{-1}$); the dispersion of subject-specific means; their variances ($\eta_{\rho_j}^{-1}$); and our uncertainty about the population mean ($\eta_\mu^{-1}$).

The variational energy of the third partition concerns the model parameters representing subject-specific latent accuracies. This energy is given by

$$I_3(\rho) = \langle \ln p(k,\mu,\lambda,\rho) \rangle_{q(\mu,\lambda)} \tag{36}$$

$$= \sum_{j=1}^m \left( k_j \ln \sigma(\rho_j) + (n_j - k_j) \ln(1 - \sigma(\rho_j)) - \frac{1}{2} a_\lambda b_\lambda (\rho_j - \mu_\mu)^2 \right) + c. \tag{37}$$

Since an analytical expression for the maximum of this energy does not exist, we resort to an iterative Newton–Raphson scheme based on a quadratic Taylor-series approximation to the variational energy $I_3(\rho)$. For this, we begin by considering the Jacobian

$$\left( \frac{dI_3(\rho)}{d\rho} \right)_j = \frac{\partial I_3(\rho)}{\partial \rho_j} = k_j - n_j \sigma(\rho_j) + a_\lambda b_\lambda (\mu_\mu - \rho) \tag{38}$$

and the Hessian

$$\left( \frac{d^2 I_3(\rho)}{d\rho^2} \right)_{jk} = \frac{\partial^2 I_3(\rho)}{\partial \rho_j \partial \rho_k} = -\delta_{jk} \left( n_j \sigma(\rho_j)(1 - \sigma(\rho_j)) + a_\lambda b_\lambda \right) \tag{39}$$

where the Kronecker delta operator $\delta_{jk}$ is 1 if $j = k$ and 0 otherwise. As noted before, the absence of off-diagonal elements in the Hessian is not based on an assumption of conditional independence of subject-specific posteriors; it is a consequence of the mean-field separation in Eq. (11). Each GN iteration performs the update

$$\rho^* \leftarrow \rho^* - \left[\frac{d^2 I_3(\rho)}{d\rho^2}\bigg|_{\rho=\rho^*}\right]^{-1} \times \frac{dI_3(\rho)}{d\rho}\bigg|_{\rho=\rho^*} \quad (40)$$

until the vector $\rho^*$ converges, i.e., $\|\rho^\square_{current} - \rho^\square_{previous}\|^2 < 10^{-3}$. Using this maximum, we can use a second-order Taylor expansion (i.e., the Laplace approximation) to set the moments of the approximate posterior:

$$\mu_\rho = \rho^* \quad \text{and} \quad (41)$$

$$\eta_\rho = -\frac{d^2 I_3(\rho)}{d\rho^2}\bigg|_{\rho=\rho^*}. \quad (42)$$

## 7. VARIATIONAL ALGORITHM AND FREE ENERGY

The expressions for the three variational energies depend on one another. This circularity can be resolved by iterating over the expressions sequentially and updating the moments of each approximate marginal given the current moments of the other marginals. This approach of conditional maximization (or stepwise ascent) maximizes the (negative) free energy $F \equiv F(q,k)$ and leads to approximate marginals that are maximally similar to the exact marginals.

The free energy itself can be expressed as the sum of the expected log-joint density (over the data and the model parameters) and the Shannon entropy of the approximate posterior:

$$F = \underbrace{\langle \ln p(k, \mu, \lambda, \rho) \rangle_q}_{\text{expected log joint}} + \underbrace{\langle -\ln q(\mu, \lambda, \rho) \rangle_q}_{\text{entropy } H[q]}. \quad (43)$$

We begin by considering the expectation of the log joint w.r.t. the variational posterior:

$$\langle \ln p(k,\mu,\lambda,\rho)\rangle_q = \sum_{j=1}^{m} \Big\langle\Big\langle \underbrace{\langle \ln \text{Bin}(k_j|\sigma(\rho_j)) + \ln N(\rho_j|\mu,\lambda)\rangle_{q(\mu)}}_{\equiv I(\rho_j)}\Big\rangle_{q(\lambda)}\Big\rangle_{q(\rho_j)}$$
$$+ \langle \ln N(\mu|\mu_0,\eta_0)\rangle_q + \langle \ln \text{Ga}(\lambda|a_0,b_0)\rangle_q.$$
(44)

The above expression contains the variational energy of $\rho j$,

$$I(\rho_j) = \ln \text{Bin}(k_j|\sigma(\rho_j)) + \frac{1}{2}(\psi(a_\lambda) + \ln b_\lambda)$$
$$- \frac{1}{2}\ln 2\pi - \frac{1}{2}a_\lambda b_\lambda \left((\rho_j - \mu_\mu)^2 + \eta_\mu^{-1}\right)$$
(45)

where $\psi(\cdot)$ is the digamma function. $I(\rho j)$ is the only term in Eq. (44) whose expectation [w.r.t. $q(\rho j)$] cannot be derived analytically. Under the Laplace approximation, however, it is replaced by a second-order Taylor expansion around the variational posterior mode $\mu_{\rho j}$,

$$I(\rho_j) \approx I(\mu_{\rho_j}) + I'(\mu_{\rho_j})(\rho_j - \mu_{\rho_j}) + \frac{1}{2}I''(\mu_{\rho_j})(\rho_j - \mu_{\rho_j})^2.$$
(46)

This allows us to approximate the expectation of $I(\rho j)$ by

$$\langle I(\rho_j)\rangle_{q(\rho_j)} \approx \underbrace{\langle I(\mu_{\rho_j})\rangle_{q(\rho_j)}}_{I(\mu_{\rho_j})} + I'(\mu_{\rho_j})\underbrace{\langle \rho_j - \mu_{\rho_j}\rangle_{q(\rho_j)}}_{0}$$
$$+ \frac{1}{2}\underbrace{I''(\mu_{\rho_j})}_{-\eta_{\rho_j}}\underbrace{\langle (\rho_j - \mu_{\rho_j})^2\rangle_{q(\rho_j)}}_{\eta_{\rho_j}^{-1}}$$
(47)
$$= I(\mu_{\rho_j}) - \frac{1}{2}$$

$$= I(\mu_{\rho_j}) - \frac{1}{2}$$
(48)

where the equality $I''(\mu_{\rho j})=-\eta_{\rho j}$ follows directly from Eq. (42). Hence, the expected log joint is:

$$\langle \ln p(k,\mu,\lambda,\rho)\rangle_q \approx \overbrace{\frac{1}{2}\ln\frac{\eta_0}{2\pi} - \frac{\eta_0}{2}\left((\mu_\mu-\mu_0)^2 + \eta_\mu^{-1}\right)}^{\langle \ln N(\mu|\mu_0,\eta_0)\rangle_q}$$

$$\overbrace{-\ln\Gamma(a_0) - a_0\ln b_0 + (a_0-1)(\psi(a_\lambda)+\ln b_\lambda) - \frac{a_\lambda b_\lambda}{b_0}}^{\langle \ln Ga(\lambda|a_0,b_0)\rangle_q}$$

$$+\sum_{j=1}^{m}\left[\ln\text{Bin}\left(k_j|\sigma(\mu_{\rho_j})\right)\right.$$

$$+\frac{1}{2}(\psi(a_\lambda)+\ln b_\lambda)-\frac{1}{2}\ln 2\pi$$

$$\left. -\frac{1}{2}a_\lambda b_\lambda\left((\mu_{\rho_j}-\mu_\mu)^2 + \eta_\mu^{-1}\right) - \frac{1}{2}\right] \quad (49)$$

The second term of the free energy in Eq. (43) is the entropy $H[q]$ of the variational posterior:

$$\langle -\ln q(\mu,\lambda,\rho)\rangle_q = \overbrace{\frac{1}{2}\ln\frac{2\pi e}{\eta_\mu}}^{H[N(\mu|\mu_\mu,\eta_\mu)]} + \sum_{j=1}^{m}\overbrace{\frac{1}{2}\ln\frac{2\pi e}{\eta_{\rho_j}}}^{H\left[N\left(\rho_j|\mu_{\rho_j},\eta_{\rho_j}\right)\right]}$$

$$+\overbrace{a_\lambda + \ln b_\lambda + \ln\Gamma(a_\lambda) + (1-a_\lambda)\psi(a_\lambda)}^{H[Ga(\lambda|a_\lambda,b_\lambda)]} \quad (50)$$

Substituting Eqs. (49) and (50) into Eq. (43) yields an expression for the free energy,

$$F \approx \frac{1}{2}\ln\frac{\eta_0}{\eta_\mu} - \frac{\eta_0}{2}\left((\mu_\mu - \mu_0)^2 + \eta_\mu^{-1}\right) + a_\lambda - a_0 \ln b_0 + \ln\frac{\Gamma(a_\lambda)}{\Gamma(a_0)}$$
$$- a_\lambda b_\lambda \left(\frac{1}{b_0} + \frac{m}{2\eta_\mu}\right) + \left(a_0 + \frac{m}{2}\right)\ln b_\lambda + \left(a_0 - a_\lambda + \frac{m}{2}\right)\psi(a_\lambda)$$
$$+ \frac{1}{2} + \sum_{j=1}^{m}\left[\ln \text{Bin}\left(k_j \big| \sigma(\mu_{\rho_j})\right) - \frac{1}{2}a_\lambda b_\lambda \left(\mu_{\rho_j} - \mu_\mu\right)^2 - \frac{1}{2}\ln \eta_{\rho_j}\right].$$
(51)

The availability of the above approximation to the free energy leads to a straightforward variational algorithm. The algorithm is initialized by setting the moments of all approximate posteriors to the moments of their respective priors. It terminates when

$$F_{current} - F_{previous} < 10^{-3} \tag{52}$$

i.e., when the free energy has converged. This criterion typically leads to the same inference as a criterion based on the parameter estimates themselves, e.g.,

$$\|\theta_{current} - \theta_{previous}\|^2 < 10^{-} \tag{53}$$

where convergence of $\theta \equiv \left(\mu_\mu, \eta_\mu, a_\lambda, b_\lambda, \mu_{\rho_1}, \ldots, \mu_{\rho_m}, \eta_{\rho_1}, \ldots, \eta_{\rho_m}\right)$ is expressed through a bound on their (squared) $l_2$-norm. However, computing (an approximation to) the free energy itself has the additional advantage that it provides an approximation to the log model evidence (see Eqs. (9) and (43)), which permits Bayesian model selection (for an example, see Brodersen et al., 2012a).

## 8. MCMC SAMPLING

The variational Bayes scheme presented above is computationally highly efficient; it typically converges after just a few iterations. However, its results are only exact to the extent to which its distributional assumptions are justified. To validate these assumptions, we compared VB to an asymptotically exact stochastic approach, i.e., Markov chain Monte Carlo (MCMC), which is computationally much more expensive than variational Bayes but exact in the limit of infinite runtime.

In the Supplemental Material, we describe a Gibbs sampler for inverting the univariate normal-binomial model introduced above. This algorithm is analogous to the one we previously introduced for the inversion of the bivariate normal-binomial model in Brodersen et al. (2012a). It proceeds by cycling over model parameters, drawing samples from their full-conditional distributions, until the desired number of samples (e.g., $10^6$) has been generated (see Supplemental Material).

Unlike VB, which was based on a mean-field assumption, the posterior obtained through MCMC retains any potential conditional dependencies among the model parameters. The algorithm is computationally burdensome; but it can be used to validate the distributional assumptions underlying variational Bayes (see Applications).

The twofold normal-binomial model for inference on the balanced accuracy

Seemingly strong classification accuracies can be trivially obtained on datasets consisting of different numbers of representatives from either class. For instance, a classifier might assign every example to the majority class and thus achieve an accuracy equal to the proportion of test cases belonging to the majority class. Thus, the use of classification accuracy as a performance measure may easily lead to optimistic inferences (Akbani et al., 2004, Brodersen et al., 2010a, Brodersen et al., 2012a, Chawla et al., 2002 and Japkowicz and Stephen, 2002).

This has motivated the use of a different performance measure: the *balanced accuracy*, defined as the arithmetic mean of sensitivity and specificity, or the average accuracy obtained on either class,

$$\varphi := \frac{1}{2}\left(\pi^+ + \pi^-\right) \tag{54}$$

where $\pi^+ := \sigma(\mu^+)$ and $\pi^- := \sigma(\mu^-)$ denote the (population) classification accuracies on positive and negative trials, respectively.[8] The balanced accuracy reduces to the conventional accuracy whenever the classifier performed equally well on either class; and it drops to chance when the classifier performed well purely because it exploited an existing class imbalance.

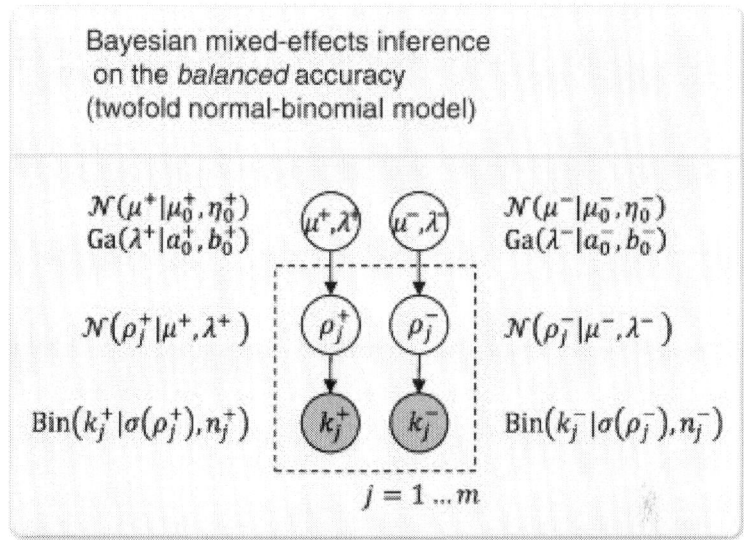

**Figure 4.** Inference on balanced accuracies. The univariate normal-binomial model (Fig. 2) can be easily extended to enable inference on the balanced accuracy. Specifically, the model is inverted separately for classification outcomes obtained on positive and negative trials. The resulting posteriors are then recombined (see main text).

We will revisit the conceptual differences between accuracies and balanced accuracies in the Discussion. In this section, we show how the univariate normal-binomial model presented above can be easily extended to allow for inference on the balanced accuracy.

We have previously explored different ways of constructing models for inference on the balanced accuracy (Brodersen et al., 2012a). Here, we infer on the balanced accuracy by duplicating our generative model for accuracies and applying it separately to data from the two classes. This constitutes the *twofold* normal-binomial model ( Fig. 4).

To infer on the balanced accuracy, we separately consider the number of correctly classified positive trials $k_j^+$ and the number of correctly predicted negative trials $k_j^-$ for each subject $j = 1\ldots m$. We next describe the true accuracies within each subject as $\pi_j^+$ and $\pi_j^-$. The population parameters $\mu^+$, $\lambda^+$ and $\mu^-$, $\lambda^-$ then represent the population accuracies on positive and negative trials, respectively.

Inverting the model proceeds by inverting its two parts independently. However, in contrast to the inversion of the *univariate* normal-binomial model, we are no longer interested in the posterior densities over the population mean accuracies $\mu^+$ and $\mu^-$ themselves. Rather, we wish to obtain the posterior density of the balanced accuracy,

$$p(\phi|k^+, k^-) = p\left(\frac{1}{2}(\sigma(\mu^+) + \sigma(\mu^-))\Big|k^+, k^-\right). \tag{55}$$

Unlike the population mean accuracy (Eq. (29)), which was logit-normally distributed, the posterior mean of the population *balanced* accuracy can no longer be expressed in closed form. The same applies to subject-specific posterior balanced accuracies. We therefore approximate the respective integrals by (one-dimensional) numerical integration. If we were interested in the *sum* of the two class-specific accuracies, $s := \sigma(\mu^+) + \sigma(\mu^-)$, we would consider the convolution of the distributions for $\sigma(\mu^+)$ and $\sigma(\mu^-)$,

$$p(s|k^+, k^-) = \int_0^s p_{\sigma(\mu^+)}(s-z|k^+) p_{\sigma(\mu^-)}(z|k^-) dz \tag{56}$$

where $p_{\sigma(\mu^+)}$ and $p_{\sigma(\mu^-)}$ represent the individual posterior distributions of the population accuracy on positive and negative trials, respectively. In the same spirit, the modified convolution

$$(\phi|k^+, k^-) = \int_0^{2\phi} p_{\sigma(\mu^+)}(2\phi-z|k^+) p_{\sigma(\mu^-)}(z|k^-) dz \tag{57}$$

yields the posterior distribution of the *arithmetic mean* of two class-specific accuracies, i.e., the balanced accuracy.

## 9. APPLICATIONS

This section illustrates the sort of inferences that can be made using VB in a classification study of a group of subjects. We begin by considering synthetic classification outcomes to evaluate the consistency of our approach and illustrate its link to classical fixed-effects and random-effects analyses. We then apply our approach to empirical fMRI data obtained from a trial-by-trial classification analysis.

## 9.1 Application to Synthetic Data

We examined the statistical properties of our approach in two typical settings: (i) a larger simulated group of subjects with many trials each; and (ii) a small group of subjects with few trials each, including missing trials. Before we turn to the results of these simulations, we will pick one simulated dataset from either setting to illustrate inferences supported by our model (Fig. 5).

The first synthetic setting is based on a group of 30 subjects with 200 trials each (i.e., 100 trials in each class). Outcomes were generated using the univariate normal-binomial model with a population mean (logit accuracy) of $\mu = 1.1$ (corresponding to a population mean accuracy of 71%) and a relatively high logit population precision of $\lambda = 4$ (corresponding to a population accuracy standard deviation of 9.3%; Fig. 5a). MCMC results were based on 100,000 samples, obtained from 8 parallel chains (see Supplemental Material).

In inverting the model, the parameter of primary interest is $\mu$, the (logit) population mean accuracy. Our simulation showed a typical result in which the posterior distribution of the population mean was sharply peaked around the true value, with its shape virtually indistinguishable from the corresponding MCMC result (Fig. 5b). In practice, a good way of summarizing the posterior is to report a central 95% posterior probability interval (or Bayesian credible interval). Although this interval is conceptually different from a classical (frequentist) 95% confidence interval, in this particular case the two intervals agreed very closely (Fig. 5c), which is typical in the context of a large sample size. In contrast, fixed-effects intervals were overconfident when based on the pooled sample accuracy and underconfident when based on the average sample accuracy (Fig. 5c).

Another informative way of summarizing the posterior population mean is to report the posterior probability mass $p$ that is below chance (e.g., 0.5 for binary classification). We refer to this probability as the (posterior) *infraliminal probability* of the classifier (cf.Brodersen et al., 2012a). Compared with a classical $p$-value, it has a deceptively similar, but more natural, interpretation. Rather than representing the frequency of observing the observed outcome (or a more extreme outcome) under the 'null' hypothesis of a classifier operating at or below chance (classical $p$-value), the infraliminal probability represents our posterior belief that the classifier does not perform better than chance. In the above simulation, we obtained $p \approx 10^{-10}$.

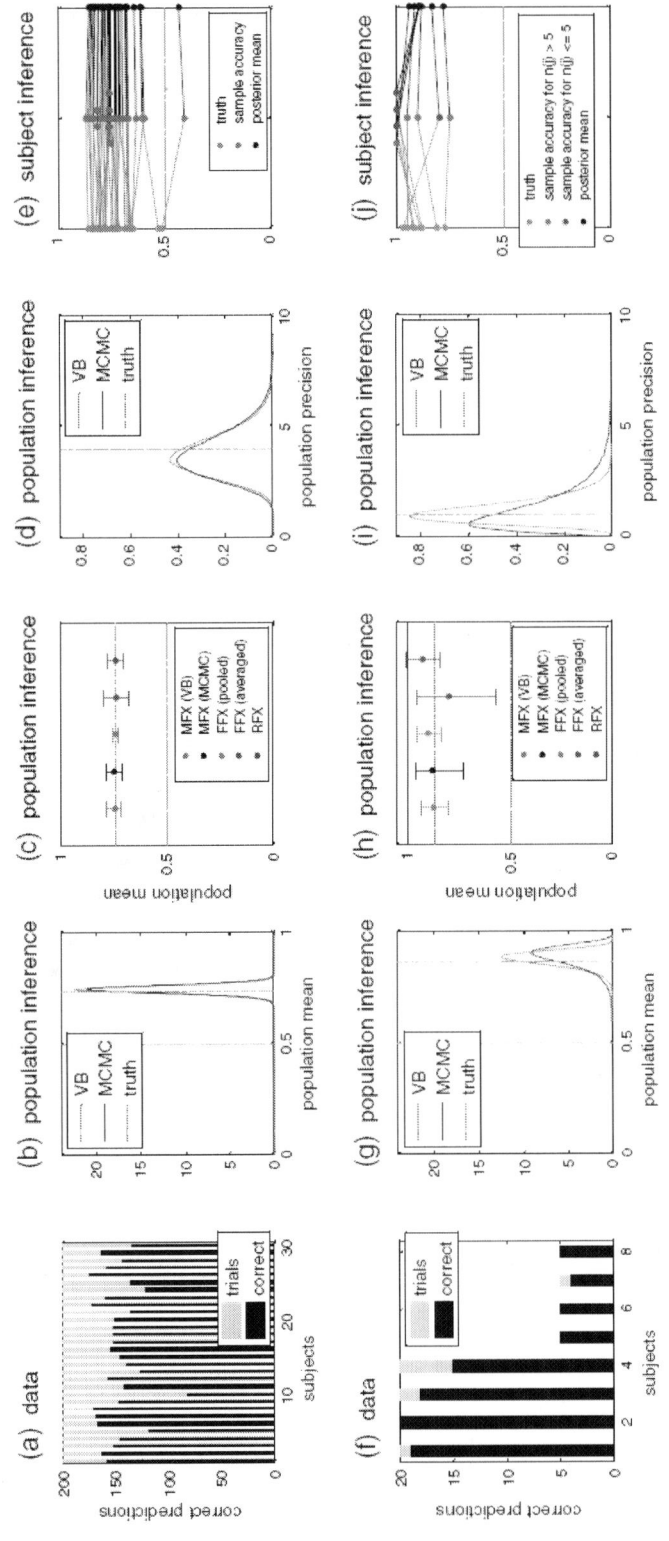

We next considered the true *subject-specific* accuracies and compared them (i) with conventional sample accuracies and (ii) with VB posterior means (Fig. 5e). This comparison highlighted one of the principal features of hierarchical models, that is, their *shrinkage* effect. Because of the limited numbers of trials, sample accuracies exhibited a larger variance than ground truth; accordingly, the posterior means, which were informed by data from the entire group, appropriately compensated for this effect by shrinking to the group mean. This effect is also known as *regression to the mean* and dates back to works as early as Galton's law of 'regression towards mediocrity' (Galton, 1886). It is obtained naturally in a hierarchical model and,

---

**Figure 5.** Application to simulated data. Two simple synthetic datasets illustrate the sort of inferences that can be obtained using a mixed-effects model. (a) Simulated data, showing the number of trials in each subject (gray) and the number of correct predictions (black). (b) Resulting posterior density of the population mean accuracy when using variational Bayes or MCMC. (c) Posterior densities can be summarized in terms of central 95% posterior intervals. Here, the two Bayesian intervals (blue/black) are compared with a frequentist random-effects 95% confidence interval and with fixed-effects intervals based on the pooled and the averaged sample accuracy. (d) Posterior densities of the population precision (inverse variance). (e) The benefits of a mixed-effects approach in subject-specific inference can be visualized (cf. Brodersen et al., 2012a) by contrasting the increase in dispersion (as we move from ground truth to sample accuracies) with the corresponding decrease in dispersion (as we move from sample accuracies to posterior means). This effect is a consequence of the hierarchical structure of the model, and it yields better estimates of ground truth (cf. Figs. 7d,h). Notably, shrinking may change the order of subjects (when sorted by accuracy) since its extent depends on the subject-specific (first-level) posterior uncertainty. Note that the x-axis does not represent any quantity by itself but simply serves to space out the three groups of data points (ground truth, samples accuracies, and posterior means). Overlapping sample accuracies are additionally scattered horizontally for better visibility. (f–j) Same plots as in the top row, but based on a different simulation setting with a much smaller number of subjects and a smaller and more heterogeneous number of trials in each subject. The smaller size of the dataset enhances the merits of mixed-effects inference over conventional approaches and increases the shrinkage effect in subject-specific accuracies.

as we will see below, leads to systematically more accurate posterior inferences at the single-subject level.

We repeated the above analysis on a sample dataset from a second simulation setting. This setting was designed to represent the example of a small group with varying numbers of trials across subjects.[9] Such a scenario is important to consider because it occurs in real-world applications whenever the number of trials eligible for subsequent classification is not entirely under experimental control. Varying numbers of trials also occur, for example, in clinical diagnostics of diseases like epilepsy where one may have different numbers of observations per patient. Classification outcomes were generated using the univariate normal-binomial model with a population mean logit accuracy of $\mu = 2.2$ and a low logit population precision of $\lambda = 1$; the corresponding population mean accuracy was 87%, with a population standard deviation of 11.2% (Fig. 5f).

Comparing the resulting posteriors (Figs. 5g–j) to those obtained on the first dataset, several differences are worth noting. Concerning the population parameters (Figs. 5g,i), all estimates remained in close agreement with ground truth; at the same time, minor discrepancies began to arise between variational and MCMC approximations, with the variational results slightly too precise (Figs. 5g,i). This can be seen best from the credible intervals (Fig. 5h, black). By comparison, an example of inappropriate inference can be seen in the frequentist confidence interval for the population accuracy, which does not only exhibit an optimistic shift towards higher performance but also includes accuracies above 100% (Fig. 5h, red).

Another typical consequence of a small dataset with variable trial numbers can be seen in the shrinkage of subject-specific inferences (Fig. 5j). In comparison to the first setting, there are fewer trials per subject, and so the shrinkage effect is stronger. In addition, subjects with fewer trials (red) are shrunk more than those with more trials (blue). Thus, the order between sample accuracies and posterior means has changed, as indicated by crossing black lines. Restoring the correct order of subjects can become important, for example, when one wishes to relate subject-specific accuracies to independent subject-specific characteristics, such as behavioral, demographic, or genetic information.

The primary advantage of VB over sampling algorithms is its computational efficiency. To illustrate this, we examined the computational load required to invert the normal-binomial model on the dataset shown in Fig. 5a. Rather than measuring computation time (which is platform-dependent), we considered the number of floating-point operations (FLOPs), which we related to the absolute error of the inferred posterior mean of

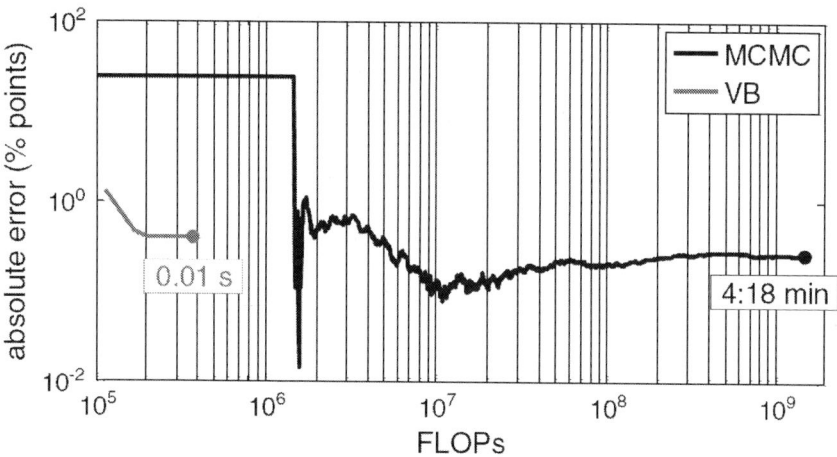

**Figure 6.** Estimation error and computational complexity. VB and MCMC differ in the way estimation error and computational complexity are traded off. The plot shows estimation error in terms of the absolute difference of the posterior mean of the population mean accuracy in percentage points (y-axis). Computational complexity is shown in terms of the number of floating point operations (FLOPs) consumed. VB converged after 370,000 FLOPs (iterative update $< 10^{-6}$) to a posterior mean of the population mean accuracy of 73.5%. Given a true population mean of 73.9%, the estimation error of VB was − 0.4 percentage points. In contrast, MCMC used up $1.47 \times 10^9$ FLOPs to draw 10,000 samples (excluding 100 burn-in samples). Its posterior mean estimate was 73.6%, implying an error of − 0.26 percentage points. Thus, while MCMC ultimately achieved a marginally lower error, VB was computationally more efficient by more than 3 orders of magnitude. It should be noted that the plot uses log–log axes for readability; the difference between the two algorithms would be visually even more striking on a linear scale.

the mean population accuracy (in percentage points; Fig. 6). We found that MCMC used 4000 times more arithmetic operations to achieve an estimate that was better than VB by no more than 0.13 percentage points.

## 9.2 Application to a Larger Number of Simulations

Moving beyond the single case examined above, we replicated our analysis many times while varying the true population mean accuracy between 0.5

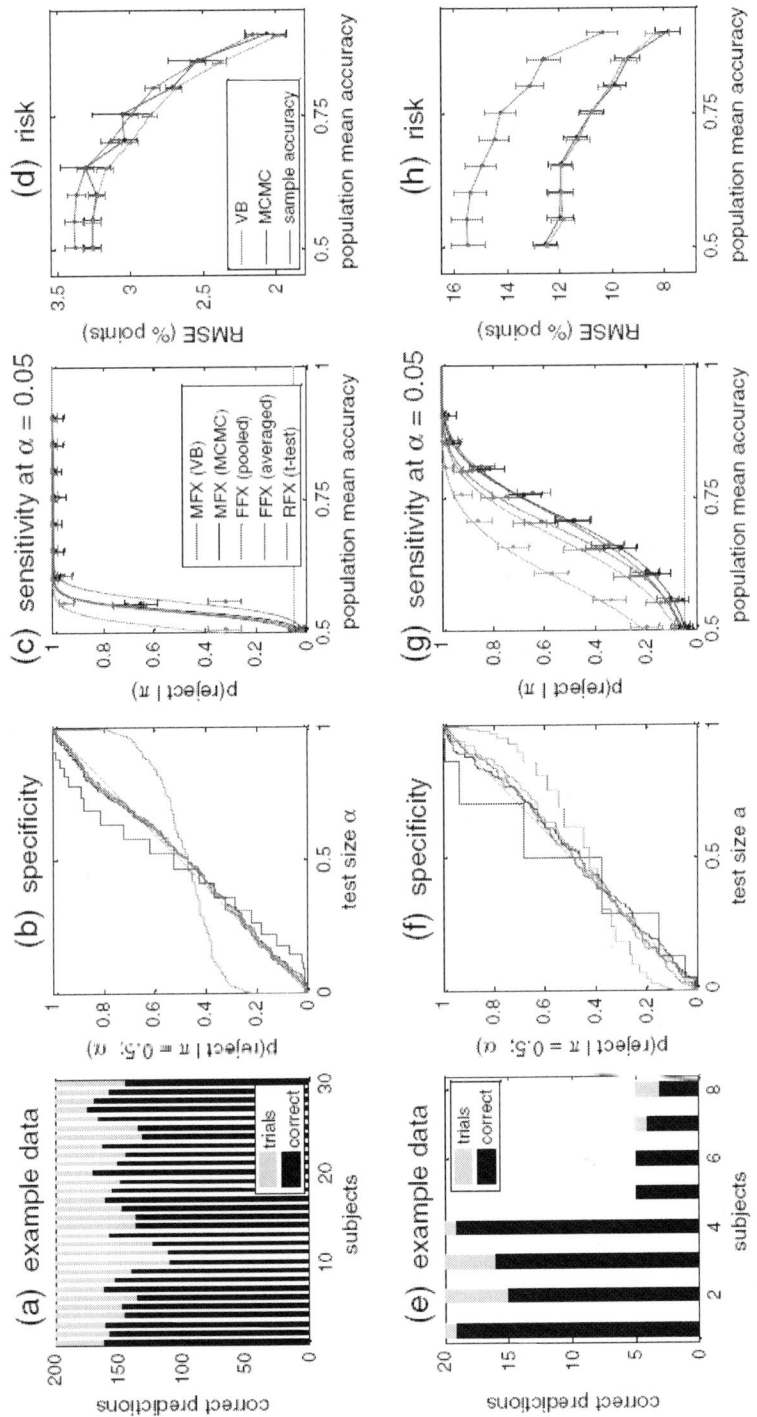

and 0.9. For each point, we ran 200 simulations. This allowed us to examine the properties of our approach from a frequentist perspective (Fig. 7).

In the first setting (Fig. 7, top row), each simulation was based on synthetic classification outcomes from 30 subjects with 200 trials each, as described in the previous section. One instance of these simulations is shown as an example (Fig. 7a); all subsequent plots are based on 200 independent datasets generated in the same way.

We began by asking, in each simulation, whether the population mean accuracy was above chance (0.5). We answered this question by computing $p$-values using the following five methods: (i) fixed-effects inference based on a binomial test on the pooled sample accuracy (orange); (ii) fixed-effects inference based on a binomial test on the average sample accuracy (violet); (iii) mixed-effects inference using VB (solid black); (iv) mixed-effects inference using an MCMC sampler with 100,000 samples (dotted black); and (v) random-effects inference using a $t$-test on subject-specific sample accuracies (red).

An important aspect of inferential conclusions (whether frequentist or Bayesian under a diffuse prior) is their validity with respect to a given test size. For example, when using a test size of $\alpha = 0.05$, we expect the test statistic to be at or beyond the corresponding critical value for the 'null' hypothesis (of the classification accuracy to be at or below the level of chance) in precisely 5% of all simulations. We thus plotted the empirical *specificity*, i.e., the fraction of false rejections, as a function of test size ( Fig. 7b). For any method to be a valid test, $p$-values should be uniformly distributed on the [0, 1] interval under the 'null'; thus, the empirical cumulative distribution function should approximate the main diagonal.

**Figure 7.** Application to a larger number of simulations. (a) One example of 200 simulations of synthetic classification outcomes (generated using the same model as in Fig. 5a). (b) Specificity of competing methods for testing whether the population mean accuracy is greater than chance, given a true population mean of 0.5. (c) Power curve, testing whether the population mean accuracy is greater than chance, given different true population mean accuracies. (d) Comparison of accuracy of subject-specific estimates, using different inference methods. (e) Example of a smaller dataset (sampled from the same model as in Fig. 5f). (f–h) Same analyses as above, but based on smaller experiments.

As can be seen from Fig. 7b, the first method violates this requirement (fixed-effects analysis, orange). It pools the data across all subjects; as a result, above-chance performance is concluded too frequently at small test sizes and not concluded frequently enough at larger test sizes. In other words, a binomial test on the pooled sample accuracy provides invalid inference on the population mean accuracy.

A second important property of inference schemes is their *sensitivity* or statistical *power* (Fig. 7c). An *ideal* test (falsely) rejects the null with a probability of $\alpha$ when the null is true, and always (correctly) rejects the null when it is false. In the presence of observation noise, such a test is only guaranteed to exist in the limit of an infinite amount of data. Thus, given a finite dataset, we can compare the power of different inference methods by examining how quickly their rejection rates rise once the null is no longer true. Using a test size of $\alpha = 0.05$, we carried out 200 simulations for each level of true population mean accuracy (0.5, 0.6, ..., 0.9) and plotted empirical rejection rates. The figure shows, as expected, that a Binomial test on the pooled sample accuracy is an invalid test, in the sense that it rejects the null hypothesis too frequently when it is true. This effect will become even clearer when using a smaller dataset (see below).[10]

Finally, we examined the performance of our VB algorithm for estimating subject-specific accuracies (Fig. 7d). We compared three estimators: (i) posterior means of $\sigma(\rho j)$ using VB; (ii) posterior means $\sigma(\rho j)$ using MCMC; and (iii) sample accuracies, i.e., individual maximum-likelihood estimates. The figure shows that posterior estimates based on a mixed-effects model led to a slightly smaller estimation error than sample accuracies. This effect was small in this scenario but became substantial when considering a smaller dataset, as described next.

In the second setting (Fig. 7, bottom row), we carried out the same analyses as above, but based on small datasets of just 8 subjects with different numbers of trials (Fig. 7e). Regarding test specificity, as before, we found fixed-effects inference to yield highly over-optimistic inferences at low test sizes (Fig. 7f).

The same picture emerged when considering sensitivities (Fig. 7g). Fixed-effects inference on the pooled sample accuracy yielded overconfident results; it systematically rejected the null hypothesis too easily. A conventional $t$-test on subject-specific sample accuracies provided a valid test, with no more false positives under the null than prescribed by the test size (red). However, it was outperformed by a mixed-effects approach (black), whose rejection probability rises more quickly when the null is no longer true, thus offering greater statistical power than the *t*-test.

Finally, in this setting of a small group size and few trials, subject-specific inference benefitted substantially from a mixed-effects model (Fig. 7h). This is due to the fact that subject-specific posteriors are informed by data from the entire group, whereas sample accuracies are only based on the data from an individual subject.

## 9.3 Accuracies versus Balanced Accuracies

As described above, the classification accuracy of an algorithm (obtained on an independent test set or through cross-validation) can be a misleading measure of generalization ability when the underlying data are not perfectly balanced. To resolve this problem, we use a straightforward extension of our model, the twofold normal-binomial model (Fig. 4), that enables inference on balanced accuracies. To illustrate the differences between the two quantities, we revisited, using our new VB algorithm, an analysis from a previous study in which we had generated an imbalanced synthetic dataset and used a linear support vector machine (SVM) for classification (Fig. 8; for details, see Brodersen et al., 2012a).

We observed that, as expected, the class imbalance caused the classifier to acquire a bias in favor of the majority class. This can be seen from the raw classification outcomes in which many more positive trials (green) than negative trials (red) were classified correctly, relative to their respective prevalence in the data (Fig. 8a). The bias is reflected accordingly by the estimated bivariate density of class-specific classification accuracies, in which the majority class consistently performed well whereas the accuracy on the minority class varied strongly, covering virtually the entire [0, 1] range (Fig. 8b). In this setting, we found that the twofold normal-binomial model of the balanced accuracy provided an excellent estimate of the true balanced accuracy under which the data had been generated (dotted green line in Fig. 8c). In stark contrast, using the single normal-binomial model to infer on the population accuracy resulted in estimates that were considerably too optimistic and therefore misleading.

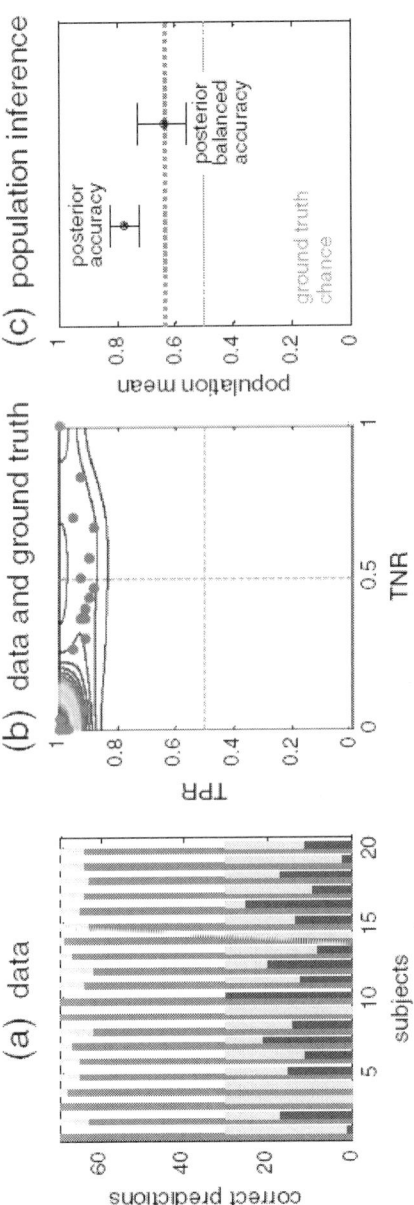

**Figure 8.** Imbalanced data and the balanced accuracy. (a) In analogy with Fig. 7a, the panel shows a set of classification outcomes obtained by applying a linear support vector machine (SVM) to synthetic data, using 5-fold cross-validation. Individual bars represent, for each subject, the number of correctly classified positive (green) and negative (red) trials, as well as the respective total number of trials (gray). (b) Sample accuracies on positive (true positive rate, TPR) and negative classes (true negative rate, TNR), based on the classification outcomes shown in (a). The underlying true population distribution is shown in terms of a bivariate Gaussian kernel density estimate (contour lines). Sample accuracies can be thought of as being drawn from this two-dimensional density; the imbalance in the data has led the SVM to acquire a bias in favor of the majority class. (c) As an example of an inference that can be obtained using the approach presented in this paper, the last panel shows central 95% posterior probability intervals of the population mean accuracy and the balanced accuracy. The plot shows that inference on the accuracy is misleading as it must be interpreted in relation to an implicit baseline that is different from 0.5. By contrast, the balanced accuracy interval provides a sharply peaked estimate of the true balanced accuracy; its baseline is 0.5.

# Variational Bayesian mixed-effects inference for classification studies

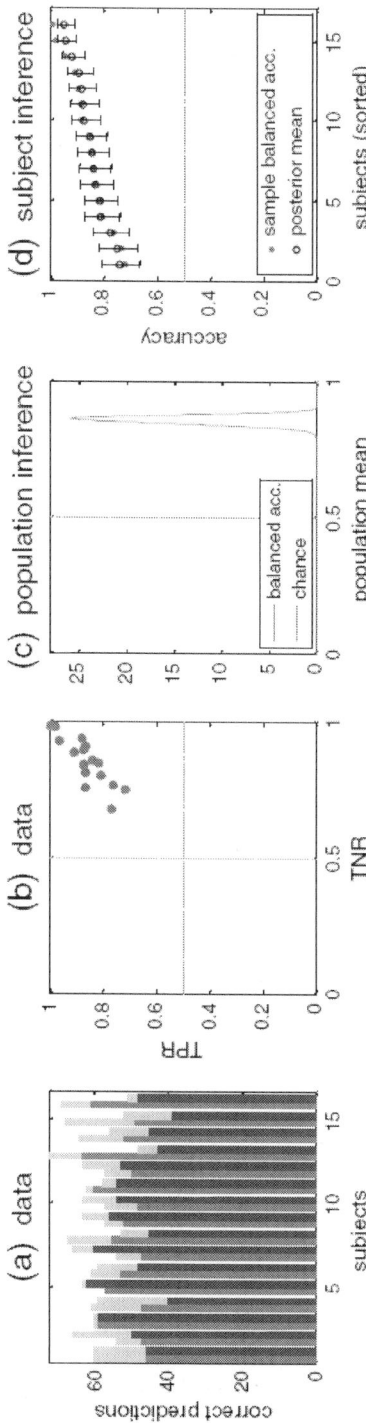

**Figure 9.** Application to empirical fMRI data: overall classification performance. (a) Classification outcomes obtained by applying a linear SVM to trial-wise fMRI data from a decision-making task. (b) Posterior population mean accuracy, inferred on using variational Bayes. (c) Posterior population precision. (d) Subject-specific posterior inferences. The plot contrasts sample accuracies with central 95% posterior probability intervals. In this case, the shrinkage effect (discrepancy between blue dots and black circles) is diminished by the large number of trials per subject.

## 9.4 Application to fMRI Data

To demonstrate the practical applicability of our VB method for mixed-effects inference, we analyzed data from an fMRI experiment involving 16 volunteers who participated in a simple decision-making task (Fig. 9). During the experiment, subjects had to choose, on each trial, between two options that were presented on the screen. Decisions were indicated by button press (left/right index finger). Details on the underlying experimental design, data acquisition, and preprocessing can be found elsewhere (Behrens et al., 2007). Here, we aimed to decode (i.e., classify) from fMRI measurements which option had been chosen on each trial. Because different choices were associated with different buttons, we expected to find highly discriminative activity in the primary motor cortex.

Separately for each subject, a general linear model (Friston et al., 1995) was used to create a set of parameter images representing trial-specific estimates of evoked brain activity in each volume element. These images entered a linear support vector machine (SVM), as implemented by Chang and Lin (2011), that was trained and tested using 5-fold cross-validation. Comparing predicted to actual choices resulted in 120 classification outcomes for each of the 16 subjects (Fig. 9a).

Using the univariate normal-binomial model for inference on the population mean accuracy, we obtained clear evidence (infraliminal probability $p < 0.001$) that the classifier was operating above chance (Fig. 9b). The variational posterior population mean balanced accuracy (posterior mean 73.7%; Fig. 9c) agreed closely with an MCMC-based posterior (73.5%; not shown). Inference on subject-specific balanced accuracies yielded fairly precise posterior intervals whose shrinkage to the population, due to the large number of trials per subject, was only small (Fig. 9d).

The overall computation time for the above VB inferences was approximately 7 ms on a 2.53 GHz Intel Xeon E5540 processor. This speedup in comparison to previous MCMC algorithms makes it feasible to construct whole-brain maps of above-chance accuracies. We illustrate this using a searchlight classification analysis (Kriegeskorte et al., 2006 and Nandy and Cordes, 2003). In this analysis, we passed a sphere (radius 6 mm) across the brain. At each location, we trained and tested a linear SVM using 5-fold cross-validation. We associated the voxel at the center of the current sphere with the number of correct predictions (i.e., the vector $k_{1:16} \in \mathbf{N}^{16}$). We then used our VB algorithm to compute a whole-brain posterior accuracy map (PAM; Fig. 10). Comprising 220,000 voxels, the map took no more than

(a) Conventional sample accuracy map (SAM) thresholded at $p < 0.001$ ($t$-tests, unc.)

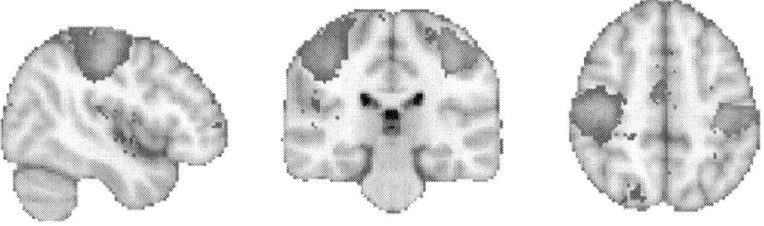

(b) Bayesian posterior accuracy map (PAM) thresholded at $p(\pi > 0.5) > 0.999$ (unc.)

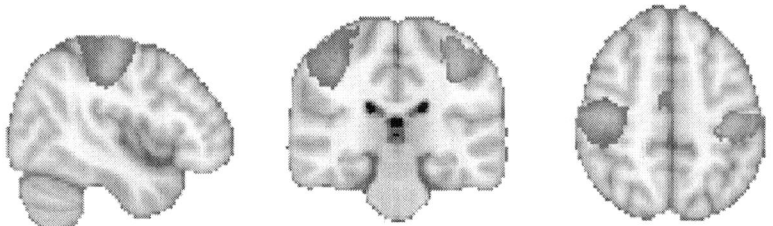

**Figure 10.** Application to empirical fMRI data: posterior accuracy map. (a) A conventional sample accuracy map (SAM) highlights regions in which a one-tailed $t$-test on subject-specific sample accuracies yielded $p < 0.001$ (uncorrected). (b) Using the VB algorithm presented in this paper, we can instead create a posterior accuracy map (PAM), which highlights those regions in which the posterior accuracy of the classification algorithm operating above chance is greater than 99.9%.

7 min 18 s to complete. The map shows the posterior population mean accuracy in voxels with an infraliminal probability of less than 0.001. Thus, it highlights regions with a posterior probability of the classifier operating above chance at the group level that is at least 99.9%.

For comparison, we contrast this result with a conventional sample accuracy map (SAM), thresholded at $p < 0.001$ (uncorrected). While the results are, overall, rather similar, the SAM shows several scattered small clusters and isolated voxels in white matter and non-motor regions that the PAM does not display.

## DISCUSSION

In this paper, we have introduced a VB algorithm for highly efficient inversion of a hierarchical Bayesian normal-binomial model which enables full mixed-effects inference on classification accuracy in group studies. Owing to its hierarchical structure, the model reflects both within-subjects and between-subjects variance components and exploits the available group data to optimally constrain inference in individual subjects. The ensuing shrinkage effects yield more accurate subject-specific estimates than those obtained through non-hierarchical models. The proposed model follows a natural parameterization and can be inverted in a fully Bayesian fashion. It is independent of the type of underlying classifier, and it supports inference both on the accuracy and on the balanced accuracy; the latter is the preferred performance measure when the data are not perfectly balanced.

In previous work, we have successfully established sampling (MCMC) approaches to Bayesian mixed-effects inference on classification accuracy at the group level (Brodersen et al., 2012a). Extending this work, the critical contribution of the present paper is the derivation and validation of a VB method for model inversion. Our new approach drastically reduces the computational complexity of previous sampling schemes (by more than 3 orders of magnitude; cf. Fig. 6) while maintaining comparable accuracy.

The parameterization used in the present manuscript differs slightly from that introduced in our previous implementations (Brodersen et al., 2012a). Specifically, we are now modeling the population distribution of subject-specific accuracies using a logit-normal density rather than a beta density. Neither is generally superior to the other; they simply represent (minimally) different assumptions about the distribution of classifier performance across subjects. However, the logit-normal density is a more natural candidate distribution when deriving a Laplace approximation, as we do here, since it implies closed-form, interpretable update equations and since it enables a straightforward approximation to the free energy (cf. Theory section). Should the issue which distribution is optimal for a given dataset at hand become a question of interest for a particular application, one can weigh the evidence for different parameterizations by means of Bayesian model selection, using the code provided in our toolbox, as shown in Brodersen et al. (2012a).

In addition to their excessive runtime, MCMC approaches to model inversion come with a range of practical challenges, such as: how to select the number of required samples; how to check for convergence, or even guarantee it; how long to design the burn-in period; how to choose the proposal distribution in Metropolis steps; how many chains to run in parallel; and how to design overdispersed initial parameter densities. By contrast, deterministic approximations such as VB involve fewer practical engineering considerations. Rather, they are based on a set of distributional assumptions that can be captured in a simple graphical model (cf. Fig. 2 and Fig. 4). While not a specific theme of this paper, it is worth reiterating that the free-energy estimate provided by VB represents an approximation to the log evidence of the model (cf. Eqs. (9) and (43)), making it easy to compare alternative distributional assumptions.

Thus, compared to previous MCMC implementations of mixed-effects inference, the present paper is fundamentally based on an idea that has been at the heart of many recent innovations in the statistical analysis of neuroimaging data: the idea that minor reductions in statistical accuracy are warranted in return for a major increase in computational efficiency.

Advances in computing power might suggest that the importance of computational efficiency should become less critical over time; but neuroimaging has repeatedly experienced how new ideas radically increase demands on computation time and thus the importance of fast algorithms. One example is provided by large-scale analyses such as searchlight approaches (Kriegeskorte et al., 2006 and Nandy and Cordes, 2003), in which we must potentially evaluate as many classification results as there are voxels in a whole-brain scan. The speed of our VB method makes it feasible to create a whole-brain map of posterior mean accuracies within a few minutes (Fig. 10). Ignoring the time taken by the classification algorithm itself, merely turning classification outcomes into posterior accuracies would have taken no less than 31 days when using an MCMC sampler with 30,000 samples for each voxel. By contrast, all computations were completed in less than 8 min when using variational Bayes, as we did in Fig. 10.

The conceptual differences between classical and Bayesian maps have been discussed extensively in the context of statistical parametric maps (SPM) and their Bayesian complements, i.e., posterior parametric maps (PPM; Friston et al., 2002 and Friston and Penny, 2003). In brief, posterior accuracy maps (PAM) confer exactly the same advantages over sample accuracy maps (SAM) as PPMs over SPMs. This makes PAMs an attractive alternative to conventional (sample-accuracy) searchlight maps.

An important feature of our approach is its flexibility with regard to performance measures. While classification algorithms used to be evaluated primarily in terms of their accuracy, the limitations of this metric have long been known and are being increasingly addressed (Akbani et al., 2004, Chawla et al., 2002 and Japkowicz and Stephen, 2002). For example, it has been suggested to restore balance by *undersampling* the larger class or by *oversampling* the smaller class. It is also possible to modify the costs of misclassification (Zhang and Lee, 2008) to prevent bias. A complementary, more generic safeguard is to replace the accuracy by the balanced accuracy, which removes the bias that may arise when a classifier is trained and tested on imbalanced data.

Fundamentally, accuracies and balanced accuracies address different scientific questions. Inference on the *accuracy* answers the question: what is the probability of making a correct prediction on a trial randomly drawn from a distribution with the same potential imbalance as that present in the current training set? Inference on the *balanced accuracy*, by contrast, answers the question: what is the probability of a correct prediction on a trial that is equally likely (a priori) to come from either class? To assess performance, this is what we are almost always interested in: the expected accuracy under a flat prior over classes.

The balanced accuracy is not confined to binary classification; it readily generalizes to $K$ classes. Specifically, the twofold normal-binomial model becomes a $K$-fold model, and the balanced accuracy $\varphi = \frac{1}{K} \sum \pi^{(k)}$ is computed on the basis of a convolution of $K$ random variables (cf. Eq. (54)).[11] Infraliminal probabilities are then determined w.r.t. the baseline level $1/K$. Even more generally, in those applications where one wishes to distinguish between different types of error (as, for example, in differential diagnostics where different misclassifications carry different costs), one could consider a weighted average of class-specific accuracies.

At first glance, another solution to dealing with imbalanced datasets would be to stick with the conventional accuracy but relate it to the correct baseline performance, i.e., the relative frequency of the majority class, rather than, e.g., 0.5 in the case of binary classification. The main weakness of this solution is that each and every report of classification performance would have to include an explicit baseline level, which would make the comparison of accuracies across studies, datasets, or classifiers slightly tedious. Future extensions of the approach presented in this paper might include functional performance measures such as the receiver-operating characteristic (ROC) or the precision-recall curve (Brodersen et al., 2010b).

Leaving classification studies aside for a moment, it is instructive to remember that mixed-effects inference and Bayesian estimation approaches have been successfully employed in other domains of neuroimaging data analysis (Fig. 11). One example are mass-univariate fMRI analyses based on the general linear model (GLM), where early fixed-effects models were

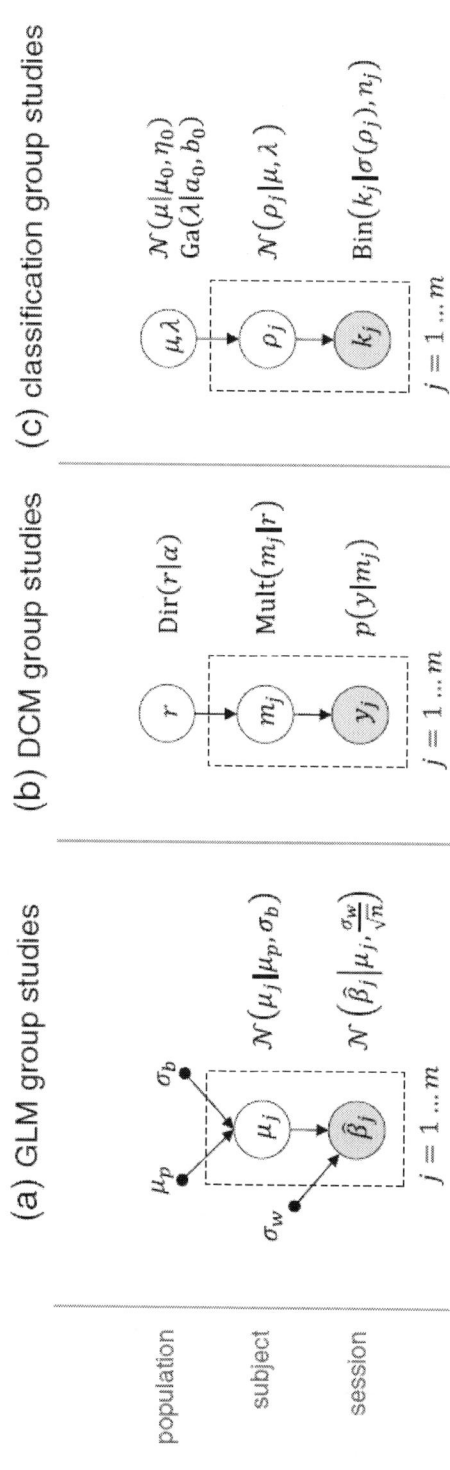

**Figure 11.** Analogies between mixed-effects models in neuroimaging. (a) The first broadly adopted models for mixed-effects inference and Bayesian estimation in neuroimaging were developed for mass-univariate fMRI analyses based on the general linear model. The figure shows a graphical representation of the (random-effects) summary-statistics approximation to mixed-effects inference. (b) Mixed-effects models have subsequently also been developed for group studies based on dynamic causal modeling (DCM). (c) The present study addresses similar issues, but in a different context, that is, in group classification analyses.

soon replaced by random-effects and full mixed-effects approaches that have since become standards in the field (Beckmann et al., 2003, Friston et al., 1999, Friston et al., 2005, Holmes and Friston, 1998, Mumford and Nichols, 2009 and Woolrich et al., 2004).

A parallel development in the domain of mass-univariate analyses has been the complementation of classical maximum-likelihood inference by Bayesian approaches (e.g., in the form of posterior probability maps; Friston and Penny, 2003). While maximum-likelihood schemes are concerned with the single most likely parameter value (i.e., mode), Bayesian inference aims for the full conditional density over possible parameter values given the data.

Another example concerns group-level model selection, e.g., in the context of dynamic causal modeling (DCM; Friston et al., 2003). Here, selecting an optimal model from a set of predefined alternatives initially rested on criteria for fixed-effects inference, such as the group Bayes factor (Stephan et al., 2007). This has subsequently been supplanted by random-effects inference that is more appropriate for typical applications in cognitive and clinical neuroscience when different mechanisms underlie measured data and thus different models are optimal across subjects (Stephan et al., 2009).

The present study addresses similar issues, but in a different context, that is, in classification group analyses. In both cases, an approximate but efficiently computable solution to a mixed-effects model (i.e., hierarchical VB) is preferable to an exact estimation of a non-hierarchical model (such as a $t$-test on sample accuracies) that disregards variability at the subject or group level. In other words: "An approximate answer to the right problem is worth a good deal more than an exact answer to an approximate problem" (John W. Tukey, 1915–2000).

It is worth noting that the model used in this paper is formally related to an earlier approach proposed by Leonard (1972). However, our choice of priors is motivated differently; we introduce a variational procedure for inference; and we use the normal-binomial model as a building block to construct larger models that can be used for inference on other performance measures, such as the balanced accuracy. Another related approach has been discussed by Olivetti et al. (2012), who carry out inference on the population mean accuracy by means of model selection between a null model and an above-chance model. For a more detailed discussion of these approaches, seeBrodersen et al. (2012a).

One assumption common to all approaches considered in this paper, whether Bayesian or frequentist, is that trial-wise classification outcomes $y_i$ are conditionally independent and identically distributed (i.i.d.) given a subject-specific accuracy $\pi j$. This assumption implies *exchangeability*, which regards the joint distribution $p(y_{1j}...y_{nj})$ as invariant to permutations of the indices $1j...nj$. Exchangeability can be safely assumed whenever no information is conveyed by the trial indices themselves (see Gelman et al., 2003, for a discussion). The stronger i.i.d. postulate is justified by assuming that test observations are conditionally i.i.d. themselves. While this may not always hold in a cross-validation setting ( Gustafsson et al., 2010, Kohavi, 1995, Pereira and Botvinick, 2011, Pereira et al., 2009 and Wickenberg-Bolin et al., 2006), it is an appropriate assumption when adopting a single-split (or hold-out) scheme, by training on one half of the data and testing on the other (cf. discussion in Brodersen et al., 2012a).

An important aspect of the proposed model is that it can be applied regardless of what underlying classifier was used; its strengths result from the fact that it accounts for the hierarchical nature of classification outcomes observed at the group level. This suggests that one might want to use classifiers that account for the data hierarchy already at the stage of classification. Unlocking the potential benefits of this approach will be an interesting theme for future work (see Gopal et al., 2012, for an example).

Finally, it is worth noting that the regularization (shrinkage) of subject-specific posterior estimates by group-level estimates which our model conveys (cf. Figs. 5j, 9d) may be beneficial for a number of real-world applications. One example are studies where the number of observations (trials) per subject cannot be controlled experimentally. This is the case in all behavioral paradigms in which the number of trials eligible for classification depends (in part) on the subject's behavior. It is also the case in some clinical applications, e.g., in epilepsy or schizophrenia. In experiments involving patients suffering from these conditions, the occurrence of epileptic and hallucinatory events, respectively, cannot be controlled by the clinician during the period of investigation. In this case, the amount by which any one subject-specific estimate is shrunk towards the population mean is correctly scaled by the number of trials in that subject (Eq. (38)). The posterior population mean, in turn, is based on the sum of these subject-specific estimates (Eq. (28)) and thus also takes into account how many trials were obtained from each subject.

In summary, the VB approach proposed in this paper is as easy to use as a $t$-test, but conveys several advantages over contemporary fixed-effects and random-effects analyses. These advantages include: (i) posterior densities as opposed to point estimates of parameters; (ii) increased sensitivity (statistical power), i.e., a higher probability of detecting a positive result, especially with small sample sizes; (iii) a shrinking-to-the-population (or regression-to-the-mean) effect whose regularization leads to more precise subject-specific accuracy estimates; and (iv) posterior accuracy maps (PAM) which provide a mixed-effects alternative to conventional sample accuracy maps (SAM).

In order to facilitate its use and dissemination, the VB approach introduced in this paper has been implemented as open-source software for both MATLAB and R. The code is freely available for download (http://www.translationalneuromodeling.org/software/). With this software we hope to assist in improving the statistical sensitivity and interpretation of results in future classification group studies.

## ACKNOWLEDGMENTS

The authors wish to thank Tom Nichols, Tim Behrens, Mark Woolrich, Adrian Groves, Ged Ridgway, and Falk Lieder for insightful discussions, and additionally Tim Behrens for sharing fMRI data. This research was supported by the University Research Priority Program 'Foundations of Human Social Behaviour' at the University of Zurich (KHB, KES), the SystemsX.ch project 'Neurochoice' (KHB, KES), and the René and Susanne Braginsky Foundation (KES).

## CONFLICT OF INTEREST

The authors declare no conflict of interest.

## REFERENCES

1. Akbani, R., Kwek, S., Japkowicz, N., 2004. Applying support vector machines to imbalanced datasets. Machine Learning: ECML 2004, pp. 39–50.

2. Attias, H., 2000. A variational Bayesian framework for graphical models. Adv. Neural Inf. Process. Syst. 12, 209–215.
3. Beckmann, C.F., Jenkinson, M., Smith, S.M., 2003. General multilevel linear modeling for group analysis in fMRI. Neuroimage 20, 1052–1063.
4. Behrens, T.E.J., Woolrich, M.W., Walton, M.E., Rushworth, M.F.S., 2007. Learning the value of information in an uncertain world. Nat. Neurosci. 10, 1214–1221.
5. Bishop, C.M., 2007. Pattern Recognition and Machine Learning. Springer, New York. Bishop, C.M., Spiegelhalter, D., Winn, J., 2002. VIBES: a variational inference engine forBayesian networks. Adv. Neural Inf. Process. Syst. 15, 777–784.
6. Blankertz, B., Lemm, S., Treder, M., Haufe, S., Müller, K.-R., 2011. Single-trial analysis and classification of ERP components—a tutorial. Neuroimage 15, 814–825.
7. Brodersen, K.H., Ong, C.S., Stephan, K.E., Buhmann, J.M., 2010a. The balanced accuracy and its posterior distribution. Proceedings of the 20th International Conference on Pattern Recognition. IEEE Computer Society, pp. 3121–3124.
8. Brodersen, K.H., Ong, C.S., Stephan, K.E., Buhmann, J.M., 2010b. The binormal assumption on precision-recall curves. Proceedings of the 20th International Conference on Pattern Recognition. IEEE Computer Society, pp. 4263–4266.
9. Brodersen, K.H., Haiss, F., Ong, C.S., Jung, F., Tittgemeyer, M., Buhmann, J.M., Weber, B., Stephan, K.E., 2011a. Model-based feature construction for multivariate decoding. Neuroimage 56, 601–615.
10. Brodersen, K.H., Schofield, T.M., Leff, A.P., Ong, C.S., Lomakina, E.I., Buhmann, J.M., Stephan, K.E., 2011b. Generative embedding for model-based classification of fMRI data. PLoS Comput. Biol. 7, e1002079.
11. Brodersen, K.H., Mathys, C., Chumbley, J.R., Daunizeau, J., Ong, C.S., Buhmann, J.M., Stephan, K.E., 2012a. Bayesian mixed-effects inference on classification performance in hierarchical data sets. J. Mach. Learn. Res. 13, 3133–3176.
12. Brodersen, K.H., Wiech, K., Lomakina, E.I., Lin, C., Buhmann, J.M., Bingel, U., Ploner, M., Stephan, K.E., Tracey, I., 2012b. Decoding the

perception of pain from fMRI using multivariate pattern analysis. Neuroimage 63, 1162–1170.

13. Chadwick, M.J., Hassabis, D., Weiskopf, N., Maguire, E.A., 2010. Decoding individual episodic memory traces in the human hippocampus. Curr. Biol. 20, 544–547.
14. Chang, C.-C., Lin, C.-J., 2011. LIBSVM: a library for support vector machines. ACM Trans.Intell. Syst. Technol. 2, 27:1–27:27.
15. Chawla, N.V., Bowyer, K.W., Hall, L.O., Kegelmeyer, W.P., 2002. SMOTE: synthetic minority over-sampling technique. J. Artif. Intell. Res. 16, 321–357.
16. Clithero, J.A., Smith, D.V., Carter, R.M., Huettel, S.A., 2011. Within- and cross-participant classifiers reveal different neural coding of information. Neuroimage 56, 699–708.
17. Cox, D.D., Savoy, R.L., 2003. Functional magnetic resonance imaging (fMRI) "brain reading": detecting and classifying distributed patterns of fMRI activity in human visual cortex. Neuroimage 19, 261–270.
18. Davatzikos, C., Resnick, S.M., Wu, X., Parmpi, P., Clark, C.M., 2008. Individual patient diagnosis of AD and FTD via high-dimensional pattern classification of MRI. Neuroimage 41, 1220–1227.
19. Dixon, P., 2008. Models of accuracy in repeated-measures designs. J. Mem. Lang. 59, 447–456.
20. Efron, B., Morris, C., 1971. Limiting the risk of Bayes and empirical Bayes estimators— part I: the Bayes case. J. Am. Stat. Assoc. 807–815.
21. Fox, C.W., Roberts, S.J., 2012. A tutorial on variational Bayesian inference. Artif. Intell.Rev. 38, 85–95.
22. Friston, K.J., Penny, W., 2003. Posterior probability maps and {SPMs}. Neuroimage 19, 1240–1249.
23. Friston, K.J., Holmes, A.P., Worsley, K.J., Poline, J.P., Frith, C.D., Frackowiak, R.S.J., 1995. Statistical parametric maps in functional imaging: a general linear approach. Hum. Brain Mapp. 2, 189–210.
24. Friston, K.J., Holmes, A.P., Worsley, K.J., 1999. How many subjects constitute a study?Neuroimage 10, 1–5.
25. Friston, K.J., Penny, W., Phillips, C., Kiebel, S., Hinton, G., Ashburner, J., 2002. Classical and Bayesian inference in neuroimaging: theory. Neuroimage 16, 465–483.

26. Friston, K.J., Harrison, L., Penny, W., 2003. Dynamic causal modelling. Neuroimage 19, 1273–1302.
27. Friston, K.J., Stephan, K.E., Lund, T.E., Morcom, A., Kiebel, S., 2005. Mixed-effects and fMRI studies. Neuroimage 24, 244–252.
28. Galton, F., 1886. Regression towards mediocrity in hereditary stature. J. Anthropol. Inst. Great Brit. Ireland 15, 246–263.
29. Gelman, A., Carlin, J.B., Stern, H.S., Rubin, D.B., 2003. Bayesian Data Analysis, 2nd ed. Chapman and Hall/CRC.
30. Ghahramani, Z., Beal, M.J., 2001. Propagation algorithms for variational Bayesian learning. Adv. Neural Inf. Process. Syst. 507–513.
31. Goldstein, H., 2010. Multilevel Statistical Models. Wiley.
32. Gopal, S., Yang, Y., Bai, B., Niculescu-Mizil, A., 2012. Bayesian models for Large-scale Hierarchical Classification. Adv. Neural Inf. Process. Syst. 25, 2420–2428.
33. Gustafsson, M.G., Wallman, M., Wickenberg Bolin, U., Göransson, H., Fryknäs, M., Andersson, C.R., Isaksson, A., 2010. Improving Bayesian credibility intervals for classifier error rates using maximum entropy empirical priors. Artif. Intell. Med. 49, 93–104.
34. Harrison, S.A., Tong, F., 2009. Decoding reveals the contents of visual working memory in early visual areas. Nature 458, 632–635.
35. 42.Hassabis, D., Chu, C., Rees, G., Weiskopf, N., Molyneux, P.D., Maguire, E.A., 2009. Decoding neuronal ensembles in the human hippocampus. Curr. Biol. 19, 546–554. Haynes, J.-D., Rees, G., 2006. Decoding mental states from brain activity in humans. Nat. Rev. Neurosci. 7, 523–534.
36. Holmes, A.P., Friston, K.J., 1998. Generalisability, random effects and population inference. Fourth Int Conf on Functional Mapping of the Human Brain. Neuroimage 7, S754.
37. Japkowicz, N., Stephen, S., 2002. The class imbalance problem: a systematic study. Intell. Data Anal. 6, 429–449.
38. Johnson, J.D., McDuff, S.G.R., Rugg, M.D., Norman, K.A., 2009. Recollection, familiarity, and cortical reinstatement: a multivoxel pattern analysis. Neuron 63, 697–708.
39. Just, M.A., Cherkassky, V.L., Aryal, S., Mitchell, T.M., 2010. A neurosemantic theory of concrete noun representation based on the underlying brain codes. PLoS One 5, e8622.

40. Klöppel, S., Stonnington, C.M., Chu, C., Draganski, B., Scahill, R.I., Rohrer, J.D., Fox, N.C., Jack, C.R., Ashburner, J., Frackowiak, R.S.J., 2008. Automatic classification of MR scans in Alzheimer's disease. Brain 131, 681–689.
41. Klöppel, S., Abdulkadir, A., Jack Jr., C.R., Koutsouleris, N., Mourão-Miranda, J., Vemuri, P., 2012. Diagnostic neuroimaging across diseases. Neuroimage 61, 457–463.
42. Knops, A., Thirion, B., Hubbard, E.M., Michel, V., Dehaene, S., 2009. Recruitment of an area involved in eye movements during mental arithmetic. Science 324, 1583–1585.
43. Kohavi, R., 1995. A study of cross-validation and bootstrap for accuracy estimation and model selection. International Joint Conference on Artificial Intelligence. Lawrence Erlbaum Associates Ltd., pp. 1137–1145.
44. Krajbich, I., Camerer, C., Ledyard, J., Rangel, A., 2009. Using neural measures of economic value to solve the public goods free-rider problem. Science 326, 596–599.
45. Kriegeskorte, N., Goebel, R., Bandettini, P., 2006. Information-based functional brain mapping. Proc. Natl. Acad. Sci. U. S. A. 103, 3863–3868.
46. Langford, J., 2005. Tutorial on practical prediction theory for classification. J. Mach. Learn. Res. 6, 273–306.
47. Lemm, S., Blankertz, B., Dickhaus, T., Müller, K.-R., 2011. Introduction to machine learning for brain imaging. Neuroimage 56, 387–399.
48. Leonard, T., 1972. Bayesian methods for binomial data. Biometrika 59, 581–589.
49. MacKay, D.J.C., 1995. Ensemble learning and evidence maximization. Proc. NIPS (Available at: http://citeseerx.ist.psu.edu/viewdoc/download?doi=10.1.1.54. 4083&rep=rep1&type=pdf [Accessed January 4, 2013]).
50. Marquand, A., Howard, M., Brammer, M., Chu, C., Coen, S., Mourão-Miranda, J., 2010. Quantitative prediction of subjective pain intensity from whole-brain fMRI data using Gaussian processes. Neuroimage 49, 2178–2189.
51. Mumford, J.A., Nichols, T., 2009. Simple group fMRI modeling and inference. Neuroimage 47, 1469–1475.

52. Nandy, R.R., Cordes, D., 2003. Novel nonparametric approach to canonical correlation analysis with applications to low CNR functional MRI data. Magn. Reson. Med. 50, 354–365.
53. Norman, K.A., Polyn, S.M., Detre, G.J., Haxby, J.V., 2006. Beyond mind-reading: multi-voxel pattern analysis of fMRI data. Trends Cogn. Sci. 10, 424–430.
54. Olivetti, E., Veeramachaneni, S., Nowakowska, E., 2012. Bayesian hypothesis testing for pattern discrimination in brain decoding. Pattern Recognit. 45, 2075–2084.
55. Penny, W.D., Stephan, K.E., Mechelli, A., Friston, K.J., 2004. Comparing dynamic causal models. Neuroimage 22, 1157–1172.
56. Pereira, F., Botvinick, M., 2011. Information mapping with pattern classifiers: a comparative study. Neuroimage 56, 476–496.
57. Pereira, F., Mitchell, T., Botvinick, M., 2009. Machine learning classifiers and fMRI: a tutorial overview. Neuroimage 45, S199–S209.
58. Schurger, A., Pereira, F., Treisman, A., Cohen, J.D., 2010. Reproducibility distinguishes conscious from nonconscious neural representations. Science 327, 97–99.
59. Sitaram, R., Weiskopf, N., Caria, A., Veit, R., Erb, M., Birbaumer, N., 2008. fMRI brain– computer interfaces: a tutorial on methods and applications. IEEE Signal Process. Mag. 25, 95–106.
60. Stelzer, J., Chen, Y., Turner, R., 2013. Statistical inference and multiple testing correction in classification-based multi-voxel pattern analysis (MVPA): random permutations and cluster size control. Neuroimage 65, 69–82.
61. Stephan, K.E., Weiskopf, N., Drysdale, P.M., Robinson, P.A., Friston, K.J., 2007. Comparing hemodynamic models with DCM. Neuroimage 38, 387–401.
62. Stephan, K.E., Penny, W.D., Daunizeau, J., Moran, R.J., Friston, K.J., 2009. Bayesian model selection for group studies. Neuroimage 46, 1004–1017.
63. Tong, F., Pratte, M.S., 2012. Decoding patterns of human brain activity. Annu. Rev. Psychol. 63, 483–509.
64. Wickenberg-Bolin, U., Goransson, H., Fryknas, M., Gustafsson, M., Isaksson, A., 2006. Improved variance estimation of classification

performance via reduction of bias caused by small sample size. BMC Bioinformatics 7, 127.
65. Woolrich, M.W., Behrens, T.E.J., Beckmann, C.F., Jenkinson, M., Smith, S.M., 2004. Multilevel linear modelling for FMRI group analysis using Bayesian inference. Neuroimage 21, 1732–1747.
66. Zhang, D., Lee, W.S., 2008. Learning classifiers without negative examples: A reduction approach. International Conference on Digital, Information Management (ICDIM).

# CHAPTER 9

## STRUCTURAL LEARNING OF BAYESIAN NETWORKS BY BACTERIAL FORAGING OPTIMIZATION

Cuicui Yang[a], Junzhong Ji[a], Jiming Liu[b,1], Jinduo Liu[a], Baocai Yin[a]

[a] College of Computer Science and Technology, Beijing University of Technology, Beijing, China

[b] Department of Computer Science, Hong Kong Baptist University, Kowloon Tong, Hong Kong

## ABSTRACT

Algorithms inspired by swarm intelligence have been used for many optimization problems and their effectiveness has been proven in many fields. We propose a new swarm intelligence algorithm for structural learning of Bayesian networks, BFO-B, based on bacterial foraging optimization. In the BFO-B algorithm, each bacterium corresponds to a candidate solution that represents a Bayesian network structure, and the algorithm operates under three principal mechanisms: chemotaxis, reproduction, and elimination and dispersal. The chemotaxis mechanism uses four operators to randomly and greedily optimize each solution in a bacterial population, then the reproduction mechanism simulates survival of the fittest to exploit superior solutions and speed convergence of the optimization. Finally, an elimination and dispersal mechanism controls the exploration processes and jumps out of a local optima with a certain probability. We tested the individual contributions of four algorithm operators and compared with two state of the art swarm intelligence based algorithms and seven other well-known algorithms on many benchmark networks. The experimental results verify that the proposed BFO-B algorithm is a viable alternative to learn the structures of Bayesian networks, and is also highly competitive compared to state of the art algorithms.

## KEYWORDS

Bacterial foraging optimization; Bayesian networks; Structural learning; Swarm intelligence

## 1. INTRODUCTION

A Bayesian network (BN) is one of the most effective theoretical models to represent uncertainty of knowledge in artificial intelligence. A BN uses a graphical model to depict conditional independence relations among random variables in a domain and encode the joint probability distribution of random variables [1]. Given a BN and observations of some variables, the values of other unobserved variables can be predicted by probabilistic inference. Therefore, systems successfully use this paradigm to model practical problems in many different areas, such as medical diagnosis, natural language processing, forecasting, biology, and control [2].

Learning a BN structure automatically from data has received much attention, and variety of learning algorithms have been proposed [3], [4], [5], [6], [7], [8], [9], [10], [11], [12], [13],[14], [15], [16], [17], [18], [19], [20], [21], [22], [23], [24], [25], [26], [27], [28] and [29]. These algorithms all adopt either the dependency analysis or score and search approaches. Dependency analysis is a constraint satisfaction problem, and employs a statistical method to judge dependency and independency relationships among variables and thereby constructs a BN [19]. Score and search is an optimization problem, and employs a search method to probe the space of BN structures and a metric to constantly evaluate each candidate network structure until the best metric value is obtained [14]. Unfortunately, both approaches have fatal drawbacks. Dependency analysis needs to perform an exponential number of dependency tests that are usually complex and unreliable, and it is hard to ensure learning quality. In contrast, learning a BN structure by score and search becomes an NP-hard problem as the number of variables increases [30]. Once the space of candidate networks becomes large, nearly all exact searches are inappropriate for BN structural learning. Although some heuristic algorithms, such as iterated local search [3], K2 [25], and hill climbing [26] and [27]algorithms can address the problem of large search spaces, they often become trapped in local optima.

To solve these problems, several stochastic algorithms based on global optimization mechanisms have been introduced for structural learning of BNs in recent years. These algorithms can be divided into two categories [31]: 1) The evolutionary algorithm, which draws inspiration from evolution and natural genetics and includes evolutionary programming, genetic algorithm, evolution strategy, and genetic programming. Evolutionary programming [4] and [5] and genetic algorithm [8], [9] and [17] based methods are effective ways with which a BN structure can be successfully learned. 2) The swarm intelligence algorithm, which is a nature inspired optimization technology that consists of particle swarm optimization (PSO) [32] and [33], ant colony optimization (ACO) [34] and [35], artificial bee colony optimization (ABC) [36], and bacterial foraging optimization (BFO) [37]. ACO, ABC, and PSO have proven their effectiveness at learning a BN structure from the data [6], [10] and [23]. Their common feature is the use of a meta-heuristic search mechanism to explore the BN structural space while a scoring metric is applied to evaluate the fitness of candidate networks.

BFO is a swarm intelligence algorithm developed by Passino in 2002 [37] and [38], which simulates the foraging behavior of *Escherichia coli* bacteria. The basic principle is that bacteria move through either tumbling or swimming to maximize the energy consumed by eating as many nutrients as they can. As the smallest creatures on earth, bacteria contain many clever optimization mechanisms. Thus, BFO has unique good performance, and has been successful in a wide variety of optimization tasks since it was proposed [39], [40], [41], [42], [43] and [44]. However, to date this optimization technology has not been applied to learning BN structures.

Existing structural learning methods PSO-B, ACO-B, and ABC-B (based on the PSO, ACO, and ABC swarm intelligence algorithms, respectively) have some latent drawbacks. PSO-B keeps track of two types of optimal solutions, which makes it easily trapped in local optima. ACO-B and ABC-B employ pheromones to construct solutions. Although the positive feedback mechanism behind the pheromone can effectively guide the search for superior solutions, if the pheromone is over used, it may overpower a better solution, and the risk of the algorithms becoming trapped in local optima is high. However, BFO does not contain mechanisms that make an algorithm easily trapped, and has a high probability of escaping from local optima. Hence, we propose a new BN structural learning method, BFO-B, based on BFO.

In BFO-B, each bacterium constantly looks for a network structure with a better metric value using three optimization mechanisms: chemotaxis, reproduction, and elimination and dispersal. The chemotaxis locally optimizes each feasible solution, reproduction applies survival of the fittest to candidate solutions, and elimination and dispersal allows jump out of a local optima. The three mechanisms maintain a balance between exploitation and exploration and make it possible to obtain a global optimal or near optimal solution. To verify BFO-B performance, we conducted a series of experiments on many benchmark networks, investigating the effects of key parameters on the algorithm performance, contributions of different mechanisms to the algorithm performance, and performance comparisons with two swarm intelligence based algorithms and seven other algorithm types. The experimental outcomes verify that BFO-B is a promising approach to learn BN structures from data, highly competitive compared with two state of the art swarm intelligence based methods, and significantly superior to other methods.

Section 2 of this paper briefly introduces BNs, the K2 scoring metric of BNs, and BFO. Section 3 presents the details of the BFO-B algorithm, and the verification experiments are described and outcomes are discussed in Section 4. Section 5 summarizes our conclusions and possible future directions.

## 2. PRELIMINARIES

### 2.1. BNs

A BN, also known as a belief network or a causal network, is a directed acyclic graph (DAG), which qualitatively characterizes the dependent and independent relationships among random variables, and uses a set of probability parameters to quantify the strength of the dependencies between each node and its parent nodes. It can be denoted as $G=(X,A)$, where $X=\{X_1, X_2, \cdots, X_i, \cdots, X_n\}$ is a set of nodes, $X_i$ is a random variable, $A=\{a_{ij}\}$ is a set of arcs, and $a_{ij}$ describes a direct dependence relationship between $X_i$ and $X_j$. A set of conditional probability parameters is also associated with each non-root node, $P(X_i|\prod(X_i))$, where $\prod(X_i)$ is a parent set of $X_i$, which quantifies how much $X_i$ depends on its parents. Thus, a BN can be uniquely encoded using the joint probability distribution of the variable set $X = \{X_1, X_2, \cdots, X_n\}$,

$$P(X_1, X_2, \cdots, X_n) = \prod_{i=1}^{n} P(X_i | \prod(X_i)). \tag{1}$$

## 2.2. K2 Scoring Metric of the BN

For the score and search approach, the problem of learning a BN can be described as follows: given a scoring metric and a training set $D = \{v_1, v_2, \ldots, v_m\}$ with $m$ cases, where each $v_i \in D$ is a fully instantiated set of $n$ random variables, a search method constantly uses the scoring metric with respect to the training set to evaluate candidate network structures, until it obtains the network structure that best matches $D$.

A key aspect of the score and search approach is the scoring metric. We use the K2 metric, one of the most well-known Bayesian scoring methods, first used in the K2 algorithm [25]. The initial expression of the K2 metric is

$$P(G:D) = P(G) \prod_{i=1}^{n} \prod_{j=1}^{q_i} \frac{(r_i - 1)!}{(N_{ij} + r_i - 1)!} \prod_{k=1}^{r_i} N_{ijk}!, \tag{2}$$

where $r_i$ is the number of possible value assignments of $X_i$, $q_i$ is the number of possible instantiations for $\prod(X_i)$, $N_{ijk}$ is the number of cases in $D$ when $X_i = v_{ik}$ and $\prod(X_i)$ are instantiated with the $j$ th configuration, and $N_{ij} = \sum_{k=1}^{r_i} N_{ijk}$.

Decomposability is a very important characteristic for a scoring metric, by which the scoring of the whole network structure can be transformed into the summed score of the local structures of each node. When a local structure of a node in a BN is changed, we only need to recalculate the scoring of this node. Thus, a decomposable scoring metric can greatly reduce the number of repeated calculations. To obtain a decomposable K2 metric, we use the logarithm of $P(G:D)$ and ignore the constant, $logP(G)$, when assuming a uniform prior for P(G) [6]. We express the decomposable K2 metric as

$$f(G:D) = \log(P(G:D)) \approx \sum_{i=1}^{n} f(X_i, \prod(X_i)), \qquad (3)$$

where $f(X_i, \prod(X_i))$ represents the K2 score of each node and is formally defined as

$$f(X_i, \prod(X_i)) = \sum_{j=1}^{q_i} \left( \log\left(\frac{(r_i-1)!}{(N_{ij}+r_i-1)!}\right) + \sum_{k=1}^{r_i} \log(N_{ijk}!) \right). \qquad (4)$$

Because the joint probability is less than 1, the decomposable K2 score using $log(P(G:D))$ is always negative. Thus, the best BN structure is that with the highest K2 score regardless of which search algorithm is used.

## 2.3. BFO

Foraging strategies are used by animals and microbes to locate, handle, and ingest food. Natural selection tends to favor those species with good foraging strategies and eliminate those with poor foraging strategies. The species with poor foraging strategies cannot obtain enough food to enable them to reproduce, so after many generations, they are either eliminated or develop good foraging strategies. Inspired by this evolutionary principle, Passino developed a BFO algorithm based on the foraging behavior of *E. coli*bacteria present in the human gut [37] and [38]. The BFO algorithm has been applied to various real-world optimization problems such as data mining [39], harmonic estimation[41], edge detection [43], and RFID network planning [44] and has shown its effectiveness.

BFO is an iteration algorithm where each bacterium represents a feasible solution of the optimization problem. It starts with a population of bacteria, randomly generated during an initialization phase. Then the bacterial population tries to find an optimal solution by three nested loop mechanisms: chemotaxis, reproduction, and elimination and dispersal. BFO combines exploitation and exploration processes. In particular, each bacterium performs exploitation processes in the chemotactic steps and controls exploration processes with a certain probability in the elimination and dispersal steps. The reproduction steps, which select superior and eliminate inferior individuals, are directly analogous with the selection mechanism of classical evolutionary algorithms and can provide fast convergence of the bacterial population near the optima [45]. The BFO algorithm is summarized in Algorithm 1, and a brief introduction to the three mechanisms is given below for the case of finding the maximum of an objective function.

**Algorithm 1** BFO.

1 Initialize population and set parameters
2 Elimination and dispersal loop: $l = l + 1$
3   Reproduction loop: $k = k + 1$
4     Chemotaxis loop: $j = j + 1$
5       Each bacterium takes a chemotactic step.
6     The bacterial population takes a reproduction step.
7   Each bacterium takes an elimination and dispersal step with a given probability.

### 2.3.1. Chemotaxis

Chemotaxis simulates the movement of *E. coli* through tumbling and swimming via flagella. A bacterium tumbles in a random direction, searching for food. If food is abundant in the selected direction, the bacterium will swim in this direction until the food supply worsens or the bacterium reaches the specified steps, which is called chemotaxis. Bacterium movement can be expressed as

$$x_i(j+1,k,l) = x_i(j,k,l) + C(i)\phi(i), \tag{5}$$

where $x_i(j,k,l)$ represents the position of the *i*th bacterium at the *j*th chemotaxis, *k*th reproduction, and *l* th elimination and dispersal step; $C(i)$ is the step size in the random direction for bacterium $i$; and $\phi(i)$ is the unit length in the random direction. When $J(x_i(j+1,k,l)) > J(x_i(j,k,l))$, the *i* th bacterium will swim another step of size $C(i)$ in the same direction according to Eq. (5). Swimming continues until the bacterium either reaches the maximum number of steps, $N_s$, or the objective function value decreases.

## 2.3.2. Reproduction

The bacteria grow longer in accordance with the increase in the absorption of nutrients in the chemotactic steps. Under appropriate conditions, some will die, and others that have obtained adequate nutrients will divide to form two daughters. To model this phenomenon, let the number of chemotactic steps, $N_c$, be the lifetime of the bacterium, $S$ be the number of the bacterial population, and $S_r = S/2$ be the number of bacteria that have accumulated adequate nutrients to cope for themselves. After $N_c$ chemotactic steps, a reproduction step is instigated. The bacterial population is sorted in descending order according to the value of a health function. A larger health value indicates a healthier bacterium, so each of the $S_r$ healthiest bacteria split into two bacteria, which are placed at the same location, while the remaining $S_r$ bacteria die, to maintain a constant population size. The health function is used to compute the accumulated objective value of the $i$th bacterium over its lifetime, and is defined as

$$J^i_{health} = \sum_{j=1}^{N_c+1} J(i,j,k,l), \qquad (6)$$

where $J^i_{health}$ represents the health value of the $i$ th bacterium, $J(i,j,k,l) = J(x_i(j,k,l))$.

## 2.3.3. Elimination and dispersal

With changes to the local environment that a population of bacteria lives in, all of the bacteria may be killed or a group of bacteria may disperse into a new environment to find better food sources. To simulate this phenomenon, an elimination and dispersal step is taken after $N_{re}$ reproduction steps. Each bacterium in the population may be eliminated or dispersed to a new location with probability $P_{ed}$. The new location is randomly initialized over the solution space, so this may or may not place bacteria near good solutions. Let $N_{ed}$ be the number of elimination and dispersal steps. Generally, $N_c \geq N_{re} \geq N_{ed}$, i.e., a bacterial population will experience many chemotactic steps before a reproduction step, and several reproduction steps before an elimination and dispersal step.

## 3. BFO FOR STRUCTURAL LEARNING OF BNS

We propose a new score and search algorithm for learning BNs using BFO. The proposed algorithm uses a bacterial population to search in the candidate network space and the K2 metric to evaluate the obtained networks until it finds the network with the highest K2 score.

### 3.1. Basic Components

To use the BFO algorithm to learn BNs, we must define some basic components.

#### 3.1.1. Representation of the problem

The solution space for learning BNs is composed of all possible DAGs. Each bacterium in the population models a feasible solution and is initialized to a DAG with fewer arcs. The bacterial population then explores in the search space to find an optimal or near optimal network structure, as identified by the K2 metric. That is, the objective function is the K2 metric Eq. (3), and the search goal is to find a network structure with the highest K2 score.

#### 3.1.2. Solution initialization

Each initial solution is generated by an iterative process. Starting from an empty graph with no arcs ($G_0$), arcs that are not already in the current graph are added one by one to the solution if and only if the score of the new solution is larger than the previous graph, and the new solution satisfies the DAG constraint. This process is repeated until the number of arcs reaches the number specified in advance. Thus, the solution initialization constructs a set of starting points in the search space. To save time, each solution is initialized to a simple graph with limited edges and subsequently optimized using the mechanisms discussed above.

252 Dynamic Programming and Bayesian Inference, Concepts and Applications

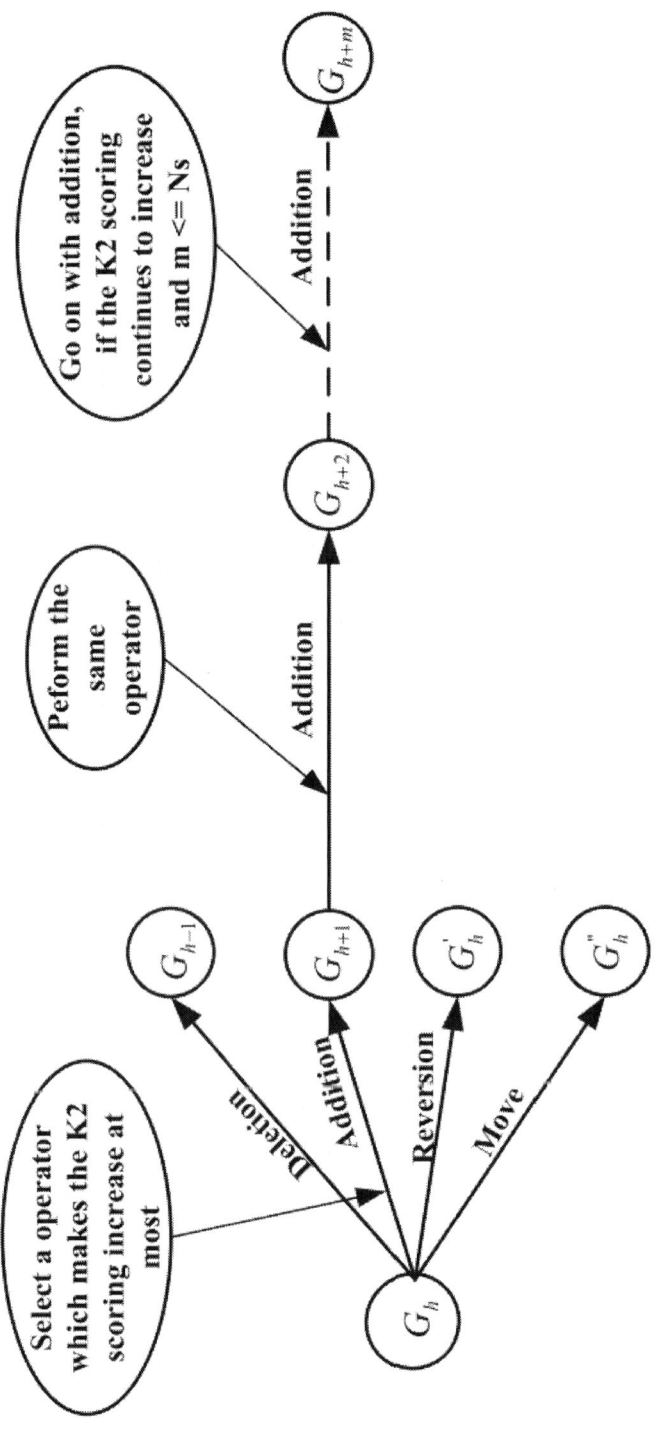

**Figure 1.** Chemotaxis steps for one bacterium.

### 3.1.3. Chemotaxis

Addition, deletion, reversion, and move operators are four candidate directions for each bacterium to select in the chemotactic process. A bacterium tries to perform each of the operators and selects the one that increases K2 score the most (equivalent to tumbling). The bacterium continues with the selected operator (equivalent to swimming) until the K2 score of the new solution no longer improves or the bacterium has performed the maximum steps, $N_s$. Essentially, the operators are four different local optimization operators. Addition, deletion, and reversion are simple standard operators in this domain and only change one edge of a candidate solution each time, which offers a relatively small range of optimization around a solution. The move operator exchanges the parent set of two existing edges in a solution and can cause a relatively large change to a solution. Thus, if a solution is not improved using the three simple operators, it may improve with the move operator. Swimming is a driving force toward a better solution using the same local optimization operator, which becomes more frequent as a bacterium approaches a better solution. Tumbling controls the change among different local optimization operators, which becomes more frequent as a bacterium moves away from a solution to search for a better one. The chemotaxis mechanism is a complex and close combination of swimming and tumbling that keeps bacteria in these places with higher scores for BN structures and plays a crucial role in searching for the best BN structures.

As shown in Fig. 1, a bacterium $G_h$, which represents a DAG with $h$ arcs, attempts deletion, addition, reversion, and move and obtains new solutions $G_{h-1}$, $G_{h+1}$, $G'_h$, and $G''_h$, respectively. Assuming the K2 score of $G_{h+1}$ is the highest, then this bacterium will pick $G_{h+1}$, and continue to test the same operator to obtain new solution, $G_{h+2}$. If the K2 score of $G_{h+2}$ is still larger than for $G_{h+1}$, it will continue to perform the same operator. This process repeats until the K2 score no longer increases or $N_s$ addition operators have been performed, i.e., $m=N_s$.

We summarize the chemotactic process of a bacterium into function **Chemotaxis_Process(*i*)**, where $K2(i,j,k,l)$ represents the score when the *i*th bacterium is at the *j*th chemotaxis, *k*th reproduction, and *l*th elimination and dispersal step:

**Chemotaxis_Process(i)**

1 Compute the K2 scoring of the graph $G(i)$ represented by bacterium $i$, noted as $K2(i, j, k, l)$, and let $G_{last} = G(i)$.
2 **Tumble:** Try to carry out the four operators on $G(i)$, respectively. Then select the one operator that makes the score of $G(i)$ increase the most, and the scoring of new graph $G(i)$ is noted as $K2(i, j + 1, k, l)$.
3 **Swim:**
   a). Let $m = 0$ (the swimming counter) and $K2_{last} = K2(i, j, k, l)$.
   b). While $m < N_s$
      i. Let $m = m + 1$
      ii. If $K2(i, j + 1, k, l) > K2_{last}$
         Let $G_{last} = G(i)$, $K2_{last} = K2(i, j + 1, k, l)$, continue the selected operator on $G(i)$, and obtain a new graph, and compute its score, $K2(i, j + 1, k, l)$.
      iii. Else, let $m = N_s$, $G(i) = G_{last}$, $K2(i, j + 1, k, l) = K2_{last}$.

## The four operators are

**Addition:** Randomly selects two nodes, $x_i$ and $x_j$, where $i \neq j$ and $x_i \notin \prod(x_j)$. If adding an arc $a_{ij} = x_i \rightarrow x_j$ in $G_h$ it does not generate a directed cycle, and a new solution $G_{h+1} = G_h \cup a_{ij}$ is obtained.

**Deletion:** Randomly selects an arc, $a_{ij}$, from node $x_i$ to $x_j$ in $G_h$, and deletes it. A new solution, $G_{h-1} = G_h \setminus a_{ij}$, is obtained.

**Reversion:** Randomly selects an arc, $a_{ij}$, from node $x_i$ to $x_j$ in $G_h$. If reversing the direction of the arc $a_{ij}$ still satisfies the DAG constraint, then $G_h = G_h \setminus a_{ij} \cup a_{ji}$.

**Move:** For two nodes $x_i$ and $x_j$ with non-empty parent sets, this operator selects a parent node for each of the two nodes, $x_k \in \prod(x_i)$ and $x_l \in \prod(x_j)$ ($k \neq l$), then exchanges $x_k$ with $x_l$ if $x_l \in (X \setminus (\prod(x_i) \cup x_i))$, $x_k \in (X \setminus (\prod(x_j) \cup x_j))$. This move operation satisfies the DAG constraint, i.e., the move operator simultaneously modifies the parent sets of two nodes.

### 3.1.4. Reproduction

Reproduction employs a health function to calculate the accumulated K2 score of each bacterium over $N_c$ chemotaxis steps, then sorts the bacterial population in descending order by their health value and picks the $S_r$ healthiest bacteria to reproduce. Each of the healthiest bacteria splits into two bacteria with the same network structures. The remaining $S_r$ least healthy bacteria are abandoned. Essentially, this mechanism provides information transmission among individuals, which is a basis to implement swarm intelligence methods. It represents a fairly abstract model of Darwinian evolution and biological genetics in genetic algorithms and acquires the elite information for delivery among swarm agents. The elite selection criterion in this mechanism is based on the fitness sum over the whole life of a bacterium, i.e., the average fitness of a bacterium in the chemotaxis process. This selection method may lead to the case where a bacterium in the last step of its life found the best solution so far, yet it still dies in the reproduction phase. However, from another perspective, it also has the effect of preventing the population from falling into local optima. Unlike the hill climbing algorithm, this selection method picks superior individuals according to multi-

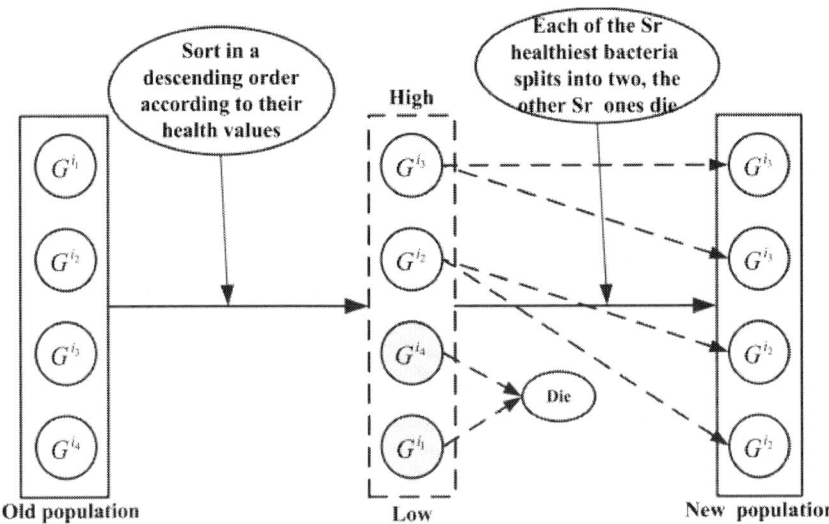

**Figure 2.** Reproduction steps for a bacterial population.

ple chemotaxis optimization processes (i.e., the average fitness) rather than picking the best every time, making it easier to escape from local optima.

The health function is

$$K2_{health}^i = \sum_{j=1}^{N_c+1} K2(i,j,k,l), \qquad (7)$$

where $K2_{health}^i$ is the health value of the $i$ th bacterium. A larger health value indicates that the corresponding bacterium has achieved larger K2 scores during its Ncchemotactic steps and hence is more likely to reproduce.

To explain the reproduction step clearly, let us take a population of four bacteria $G^1_1$, $G^1_2$, $G^1_3$, and $G^1_4$ as an example. The process is shown in Fig. 2. First, the bacteria are sorted in descending order on the basis of their health value, obtaining $G^i_3$, $G^i_2$, $G^i_4$, and $G^i_1$, i.e., $G^i_3$ and $G^i_2$ are the healthiest bacteria, $G^i_4$ and $G^i_1$ the least healthy. $G^i_4$ and $G^i_1$ are discarded, and $G^i_3$ and $G^i_2$ reproduce themselves to create two $G^i_3$ and two $G^i_2$.

### 3.1.5. Elimination and dispersal

The elimination and dispersal mechanism is invoked after $N_{re}$ reproduction steps. Each bacterium in the population is subjected to an elimination and dispersal step with probability $P_{ed}$ ($0 < P_{ed} < 1$) and the rule is

$$G = \begin{cases} G' & \text{if } q < P_{ed} \\ G & \text{otherwise} \end{cases}, \tag{8}$$

where $q$ is a random number uniformly distributed in $[0,1]$, $G$ is the current solution associated with a bacterium and $G'$ is a new solution obtained by solution initialization. For each bacterium, if the random number is smaller than $P_{ed}$, it moves to a new initial solution; otherwise, the solution remains unchanged. This mechanism generates new solutions for some bacteria and makes these bacteria search from new starting points, which ensures new search regions are explored across the search space. Thus, the mechanism helps the bacterial population jump out of local optima.

## 3.2. Algorithm Description

The proposed BFO-B algorithm is shown in Algorithm 2. It starts with an initial population of DAGs, randomly generated by solution initialization, and iteratively performs the three principal mechanisms (chemotaxis, reproduction, and elimination and dispersal) to search for networks with higher scores. BFO-B uses chemotaxis and elimination and dispersal to balance between exploitation and exploration. It performs exploitation processes to local solutions in the chemotaxis phase and exploration processes in the elimination and dispersal phase. Chemotaxis is a major driving force that provides local optimization, where the bacteria try attain larger K2 scores, and can be viewed as a biased random walk or stochastic hill climbing. Elimination and dispersal places some bacteria in new regions, which causes a certain amount of destruction of the accumulating chemotactic progress, but it also has the positive effect of assisting chemotaxis because the dispersal may help bacteria jump out of local optimal solutions and obtain a global optima. Therefore, elimination and dispersal is consider to be a chemotactic mobile behavior at the population level. Reproduction provides information transmission among the whole population and picks the elite individuals with higher scores. This process can accelerate convergence, and the special selection method based on average fitness will, to some extent, avoid the algorithm converging to the local best solutions.

**Algorithm 2** BFO-B.

**Input:** D (Dataset)
**Output:** Bayesian network

1 **Initialization:**
   **a). Set parameters:** $S, N_s, N_c, N_{re}, N_{ed}, P_{ed}, S_r$
   $S$: population size of the bacterial colony.
   $N_s$: maximum number of swimming loops,
   $N_c$: maximum number of chemotaxis loops,
   $N_{re}$: maximum number of reproduction loops,
   $N_{ed}$: maximum number of elimination-and-dispersal loops,
   $P_{ed}$: the probability of each bacterium performing elimination and dispersal.
   $S_r$: the number of high-health bacteria that would copy themselves in reproduction process.
   **b). Initialize the bacterial population**
   For $i = 1$ to $S$
      do the process of solution initialization, and get the $i$th graph $G(i)$ represented by bacterium $i$.
   Let the best graph $G_{best} = G(1)$.
   **c). Let j = k = l = 0 (three counters)**
2 **Elimination and dispersal loop:** $l = l + 1$
3   **Reproduction loop:** $k = k + 1$
4     **Chemotaxis loop:** $j = j + 1$
      **For bacterium** $i = 1, 2, ..., S$
         Perform **Chemotaxis_Process(i)**
         if $(f(G(i) : D) > f(G_{best} : D))$ (i.e., if the scoring of $G(i)$ is higher than that of $G_{best}$)
            $G_{best} = G(i)$.
5     If $j < N_c$, go to step 4.
6   **Reproduction:**
    **a).** For given $k, l$, compute the health values of the current population according to Eq. (7).
    **b).** Sort bacteria in order of descending $J_{health}$.
    **c).** Split each of the $S_r$ healthiest bacteria into two new ones while abandoning the other $S_r$ ones.
7   If $k < N_{re}$, go to step 3.
8   **Elimination and dispersal:**
    For each bacterium $i = 1, 2, ..., S$ may eliminate and disperse with a probability $P_{ed}$ according to Eq. (8).
9 If $l < N_{ed}$, go to step 2, otherwise end.
10 **Return** $G_{best}$.

## 4. EXPERIMENTAL EVALUATION

We conducted experiments to study the performance of the BFO-B algorithm and compared it with nine well-known algorithms on many benchmark networks.

## 4.1. Experimental Methodology

An algorithm for learning BNs is generally evaluated by testing it on datasets generated from benchmark networks by probabilistic logic sampling. In our experiments, we use twelve benchmark networks of different sizes, as detailed in Table 1, including their source and domain descriptions. Alarm is the most common network used in the literature for learning BN structures from data. Insurance, Child, and Asia are also relatively common networks. We also used two tiled networks, Alarm3 and Child3, which are separately composed of two common networks, Alarm and Child, by tiling three copies of themselves [15]. The tiling is performed in a way that maintains the structural and probabilistic properties of the original network in the tiled network. The aim of using tiled networks is to check the performance of an algorithm as the number of variables increases, while the difficulty of learning the network remains the same. All of the datasets used in the experiments are generated from the networks by probabilistic logic sampling, as shown in Table 2, where the datasets (D), the original BNs (G) that generate the datasets, the number of cases in D, the number of nodes in G, the number of arcs in G, and the K2 scores for the true network structures are listed for reference. Some datasets originate from the same network but have different data volume. For example, eight datasets are generated from the most common Alarm network. Using datasets from the same network, we can check the algorithm performance as data volume increases. The experimental platform was a PC with Core 2, 2.13 GHz CPU, 2.99 GB RAM, and Windows XP. The proposed BFO-B algorithm was implemented in the Java language.

**Table 1.** Benchmark networks used

| ID | Network[a] | Domain description |
|---|---|---|
| 1 | Alarm[a] | A network by medical experts for monitoring patients in intensive care. |
| 2 | Insurance[a] | A network for evaluating car insurance risks. |
| 3 | Child[a] | A preliminary diagnostic model for newborn babies with congenital heart disease. |
| 4 | Credit[a] | A model for assessing worthiness of an individual. |
| 5 | Asia[a] | A fictitious medical example on whether a patient has tuberculosis. |
| 6 | EngineFuelSystem[a] | A diagnostic model for a vehicle fuel system. |
| 7 | Boerlage92[b] | A model for the relationships between the beliefs of Jim, a fictitious character who lives in a small community that also contains Hank, Tom, Molly, and Gale. |
| 8 | Studfarm[c] | A model for calculating the probabilities of the horses in a stud farm being carriers of a recessive gene causing a life threatening disease. |
| 9 | Brain[c] | A model of a brain network used in medicine for diagnosing mental illness. |
| 10 | Tank[a] | A diagnosis model for possible explosion in a tank. |
| 11 | Win95pts[a] | An expert system for printer troubleshooting in Windows 95. |
| 12 | Hepar II[a] | A diagnostic model for liver damage. |

[a] Available in the software GeNie or https://dslpitt.org/genie/.
[b] Available in the software Hugin 8.2 (×64).
[c] Available from http://www.fmrib.ox.ac.uk/analysis/netsim/index.html.

**Table 2.** Datasets used.

| Dataset (D) | Network (G) | Num. of cases | Num. of nodes | Num. of arcs | Scoring |
|---|---|---|---|---|---|
| Alarm-1000 | Alarm | 1000 | 37 | 46 | −5034.53 |
| Alarm-2000 | Alarm | 2000 | 37 | 46 | −9729.13 |
| Alarm-3000 | Alarm | 3000 | 37 | 46 | −14412.69 |
| Alarm-4000 | Alarm | 4000 | 37 | 46 | −19110.77 |
| Alarm-5000 | Alarm | 5000 | 37 | 46 | −23793.81 |
| Alarm-6000 | Alarm | 6000 | 37 | 46 | −28358.21 |
| Alarm-7000 | Alarm | 7000 | 37 | 46 | −33033.05 |
| Alarm-8000 | Alarm | 8000 | 37 | 46 | −37755.72 |
| Insurance-3000 | Insurance | 3000 | 27 | 52 | −19843.05 |
| Insurance-5000 | Insurance | 5000 | 27 | 52 | −32284.47 |
| Insurance-6000 | Insurance | 6000 | 27 | 52 | −38466.62 |
| Child-2000 | Child | 2000 | 20 | 25 | −10718.35 |
| Child-5000 | Child | 5000 | 20 | 25 | −26577.46 |
| Credit-3000 | Credit | 3000 | 12 | 12 | −13844.74 |
| Credit-6000 | Credit | 6000 | 12 | 12 | −27550.88 |
| Asia-1000 | Asia | 1000 | 8 | 8 | −1013.21 |
| Asia-5000 | Asia | 5000 | 8 | 8 | −4884.05 |
| Child3-5000 | Child3 | 5000 | 60 | 79 | −81785.38 |
| Alarm3-5000 | Alarm3 | 5000 | 111 | 149 | −75355.59 |
| EngineFuelSystem-10000 | EngineFuelSystem | 10000 | 9 | 11 | −2272.03 |
| Boerlage92-10000 | Boerlage92 | 10000 | 23 | 36 | −44227.86 |
| Studfarm-10000 | Studfarm | 10000 | 12 | 14 | −1662.17 |
| Brain-10000 | Brain | 10000 | 50 | 66 | −290849.05 |
| Tank-10000 | Tank | 10000 | 14 | 20 | −14253.76 |
| Win95pts-50000 | Win95pts | 50000 | 76 | 112 | −196846.98 |
| Hepar II-50000 | Hepar II | 50000 | 70 | 123 | −705617.38 |

The score and search approach can evaluate the learned results based on the scores (e.g. the K2 metric), and/or structural difference, i.e., the number of different arcs in the learned network compared with the true network. The higher the K2 score or the smaller the structural difference implies a better outcome. However, the network with the highest K2 score is not necessarily the same network that has the smallest structural difference. Here, we are primarily concerned with researching search algorithms, so we consider the K2 metric as the primary way to evaluate the learned results.

In the experiments, we first analyzed the BFO-B parameter influences by empirical testing and chose appropriate values for each parameter. To probe the BFO-B algorithm details, we investigated the contributions of different mechanisms to algorithm performance. BFO-B was then run on all the datasets in Table 2 to test its performance. Finally, using various metrics, we compared BFO-B with nine well-known methods, including six score and search algorithms, two dependency analysis algorithms, and a hybrid method combing score and search and dependency analysis.

We used statistical analysis [46] to provide levels of confidence in our comparisons. Kolmogorov–Smirnov tests showed that our results do not follow a Gaussian distribution. Therefore, we used Kruskal–Wallis tests to perform nonparametric analysis of the results. The Friedman test also provides nonparametric analysis between more than two samples. However, the Friedman test is only valid for correlated samples, whereas the Kruskal–Wallis test may be applied to independent samples. In the experiments, the BFO-B algorithm and each comparison algorithm were independently run, so there is no correlation. Hence, we selected the Kruskal–Wallis test. To demonstrate the differences between the proposed algorithm and any of the comparison algorithms, we present all the statistical results whether the difference is significant or not. We apply the Kruskal–Wallis test to perform paired tests with the confidence level 95%, i.e., the probability of producing the difference by chance is not greater than 5%. If the $p$-value obtained from the test is less than 5%, we consider that a significant difference exists in the corresponding experimental results.

## 4.2. Main Parameter Selection

There are six parameters, $S$, $N_s$, $N_c$, $N_{re}$, $N_{ed}$, and $P_{ed}$ for the bacterial population size, number of swimming steps, number of chemotactic steps, number of reproduction steps, number of elimination and dispersal steps, and the elimination and dispersal probability, respectively. Fortunately, some BFO-B parameters are easily determined. For example, the number of swimming along the same operator in the chemotaxis phase is always less than five, and the algorithm performance is best when $N_s = 4$ (from test experiments). Since $N_c \geq N_{re} \geq N_{ed}$, $N_{re}$ and $N_{ed}$ also have constrained values. Thus, for our experiments, we set $N_s = 4$, $N_{re} = 4$, and $N_{ed} = 3$.

In general, swarm intelligence based algorithms have good robustness, and are less sensitive to the parameters. Hence, we select only one dataset (Alarm-2000), which is generated from the most common Alarm network, as an example of the process of determining the remaining three parameters: $S$, $N_{ed}$, and $P_{ed}$. Ten candidate values of each parameter were chosen. The value of each parameter was changed while keeping the other parameters fixed and 10 independent trials performed for each set of parameters. The evaluation metrics used are

HKS: the Highest K2 Score over all trials. The higher the corresponding value, the better the learned network.

SSS: the K2 Scoring of the learned network with the Smallest Structural difference over all trials.

LKS: the Lowest K2 Score over all trials. The higher the corresponding mean value, the better the solution performance of the algorithm.

AKS: the Average K2 Score over all trials. The higher the corresponding mean value, the better the solution performance of the algorithm.

AET: the Average Execution Time over all trials. The smaller the corresponding mean value, the better the time performance of the algorithm.

**(A) Size of the bacterial population ($S$)** The size of a population plays an important role in the search process of a swarm intelligence algorithm. To probe the effect of this parameter on our algorithm performance and select an appropriate value, we performed BFO-B 10 times using each candidate value of $S$, as summarized in Table 3. AKS for $S \geq 80$ is significantly superior to $S < 80$. Moreover, HKS, SSS, and LKS show that BFO-B can not only consistently find networks with the highest K2 scores ($-9717.46$), but also all the scores are between the highest score ($-9717.46$) and the score of the network with the smallest structural difference ($-9720.09$) when $S \geq 80$. Thus, larger $S$ is better for the algorithm, because when there are more bacteria to search for the best network in a search space, it is more likely that the

**Table 3.** BFO-B for different $S$ on Alarm-2000.

| $S$ | HKS[a] | SSS[a] | LKS | AKS[b] | AET[b] (s) |
|---|---|---|---|---|---|
| 30 | −9717.46(4) | −9720.19(1) | −9753.19 | −9725.16 ± 10.37 | 120.5 ± 9.6 |
| 40 | −9717.46(4) | −9717.55(1) | −9728.15 | −9719.77 ± 3.64 | 148.7 ± 18.7 |
| 50 | −9717.46(6) | −9720.09(1) | −9725.15 | −9718.76 ± 2.30 | 182.9 ± 19.3 |
| 60 | −9717.46(6) | −9720.09(1) | −9734.29 | −9720.78 ± 5.24 | 188.1 ± 9.0 |
| 70 | −9717.46(5) | −9720.09(2) | −9722.41 | −9718.78 ± 1.67 | 232.5 ± 33.7 |
| 80 | **−9717.46(8)** | −9720.09(1) | **−9720.09** | −9717.96 ± 0.99 | **251.6 ± 28.9** |
| 90 | **−9717.46(8)** | −9720.09(1) | **−9720.09** | −9717.73 ± 0.78 | 311.5 ± 52.6 |
| 100 | **−9717.46(6)** | −9720.09(1) | **−9720.09** | −9718.20 ± 1.10 | 310.2 ± 31.9 |
| 110 | **−9717.46(9)** | −9720.09(1) | **−9720.09** | −9717.73 ± 0.79 | 319.2 ± 30.1 |
| 120 | **−9717.46(9)** | −9717.46(9) | **−9717.82** | −9717.50 ± 0.11 | 345.6 ± 43.9 |

a The numbers in parentheses indicate the number to obtain the corresponding K2 scores.
b Results in the form $\mu \pm \delta$ indicate the mean $\mu$ and the standard deviation $\delta$.

**Table 4.** Significance (*p*-values [a] from Kruskal–Wallis tests) for *S*.

| S   | 80 | 90     | 100    | 110    | 120    |
|-----|----|--------|--------|--------|--------|
| 80  |    | 0.5839 | 0.4269 | 0.5839 | 0.4656 |
| 90  |    |        | 0.1793 | 1      | 0.9422 |
| 100 |    |        |        | 0.1793 | 0.1117 |
| 110 |    |        |        |        | 0.9422 |
| 120 |    |        |        |        |        |

a The confidence level is 95%.

optimal or near optimal network structure will be found. To differentiate the most appropriate value from candidate values $S \geq 80$, i.e., 80, 90, 100, 110, and 120, we performed pair-wise Kruskal–Wallis tests on the learning results, as shown in Table 4, in the form of *p*-values, where $p > 0.05$ indicates that the two values of *S* have no statistically significant different effect on the performance and vice versa. All *p*-values are greater than 0.05, indicating that there are no statistically significant differences between the five values. However, running time increases as *S* increases, as shown in Table 3, and so we would prefer a smaller value. Thus, balancing between finding a better network and spending less time, we set size of the bacterial population to 80.

**(B) Chemotactic steps ($N_c$)** Chemotaxis is a key process, and so $N_c$ is an important parameter for BFO-B. We tested BFO-B with 10 different candidate values of $N_c$, as shown in Table 5. From HKS, the BFO-B algorithm generally finds the network with the highest score (−9717.46) when $N_c \geq 20$; SSS shows that BFO-B finds the network with the smallest structural difference (−9720.09) when $N_c = 25, 30$, or $\geq 40$; and LKS shows that BFO-B obtains the highest value (−9720.09) when $N_c = 30, 45$, and 60. Thus, the BFO-B algorithm shows good performance when $N_c = 30, 45$, or 60 for all three statistics. To identify the most appropriate value, we selected five candidate values of 30, 40, 45, 50, and 60 and performed pair-wise Kruskal–Wallis tests, as shown in Table 6. All *p*-values are greater than 0.05, indicating there are no statistically significant differences between the five values. However, AET shows that running time generally increases as $N_c$ increases, and so we set $N_c = 30$.

**(C) Elimination and dispersal probability ($P_{ed}$)** $P_{ed}$ controls the exploratory search in the elimination and dispersal process. To select an appropriate value for $P_{ed}$, we tested BFO-B with 10 different candidate values, as shown

**Table 5.** BFO-B for different $N_c$ on Alarm-2000.

| $N_c$ | HKS[a] | SSS[a] | LKS | AKS[b] | AET[b] (s) |
|---|---|---|---|---|---|
| 15 | −9729.29(1) | −9729.08(1) | −9815.46 | −9755.08 ± 26.47 | 170.6 ± 6.3 |
| 20 | −9717.46(4) | −9719.03(1) | −9732.21 | −9721.37 ± 5.63 | 209.6 ± 12.0 |
| 25 | −9717.46(7) | −9720.09(2) | −9726.39 | −9718.88 ± 2.71 | 226.9 ± 21.1 |
| **30** | −9717.46(8) | −9720.09(1) | −9720.09 | −9717.96 ± 0.99 | **251.6 ± 28.9** |
| 35 | −9717.46(5) | −9717.46(5) | −9725.53 | −9719.49 ± 2.81 | 254.1 ± 37.1 |
| **40** | −9717.46(7) | −9720.09(1) | −9720.94 | −9718.08 ± 1.23 | 266.2 ± 34.7 |
| **45** | −9717.46(9) | −9720.09(1) | −9720.09 | −9717.73 ± 0.79 | 274.7 ± 30.5 |
| **50** | −9717.46(7) | −9720.09(1) | −9723.42 | −9718.58 ± 1.92 | 283.6 ± 45.6 |
| 55 | −9717.46(6) | −9720.09(2) | −9724.95 | −9719.23 ± 2.51 | 277.3 ± 33.6 |
| **60** | −9717.46(7) | −9720.09(3) | −9720.09 | −9718.25 ± 1.20 | 257.8 ± 25.7 |

a The numbers in parentheses indicate the number to obtain the corresponding K2 scores.
b Results in the form $\mu \pm \delta$ indicate the mean $\mu$ and the standard deviation $\delta$.

**Table 6.** Significance (p-values [a] for Kruskal–Wallis tests) for $N_c$.

| $N_c$ | 30 | 40 | 45 | 50 | 60 |
|---|---|---|---|---|---|
| 30 |    | 0.2792 | 1      | 0.2548 | 0.2758 |
| 40 |    |        | 0.2792 | 0.8886 | 1      |
| 45 |    |        |        | 0.2548 | 0.2758 |
| 50 |    |        |        |        | 0.8857 |
| 60 |    |        |        |        |        |

a The confidence level is 95%.

in Table 7. AKS shows that the performance when $P_{ed}=0.10, 0.12, 0.14$, and 0.15 is superior to the other cases. HKS, SSS, and LKS show that the BFO-B algorithm not only finds the network with the highest score (−9717.46) and that with the smallest structural difference (−9720.09), but also all of the scores of the 10 runs are between the highest score (−9717.46) and the score of the network with the smallest structural difference (−9720.09) for those $P_{ed}$ cases. Although $P_{ed}=0.18$ returned 9 of 10 runs with the highest K2 score (−9717.46), which is better than the four cases above, the average score is inferior. Thus, $P_{ed}$ either too large or too small adversely affects BFO-B performance. If $P_{ed}$ is too small, there will be fewer bacteria to perform the elimination and dispersal step, making it more difficult to overcome a local optima. On the other hand, if $P_{ed}$ is too large, while more bacteria will restart to search for the optimal solution, there is also increasing risk of missing the optimal solution. To further select the most appropriate value from the five candidates (0.10, 0.12, 0.14, 0.15, and 0.18), we performed pair-wise Kruskal–Wallis tests, as shown in Table 8. All $p$-values are greater than 0.05, indicating that there are no statistically significant differences between the five candidate values. Since $P_{ed}=0.10$ shows the best time performance among the five cases, we selected $P_{ed}=0.10$.

Summarizing, the BFO-B parameters used in our experiments were: $N_s=4$, $N_{re}=4$, $N_{ed}=3$, $S=80$, $N_c=30$, and $P_{ed}=0.1$.

## 4.3. Effects of Different Components

The proposed BFO-B algorithm combines several different components that work together to obtain the final outcomes. To gain additional insight into the inner workings of the algorithm, we probed individual contributions of the different components, i.e., the inherent characteristics of BFO-B. Three datasets generated from different networks were chosen to validate the effects of different components. When choosing the datasets, we consid-

**Table 7.** BFO-B for different $P_{ed}$ on Alarm-2000.

| $P_{ed}$ | HKS[a] | SSS[a] | LKS | AKS[b] | AET[b] (s) |
|---|---|---|---|---|---|
| 0.05 | −9717.46(4) | −9720.09(2) | −9725.53 | −9719.39 ± 2.60 | 240.7 ± 33.4 |
| 0.08 | −9717.46(4) | −9720.09(2) | −9726.44 | −9720.18 ± 3.31 | 246.0 ± 30.8 |
| **0.10** | **−9717.46(8)** | **−9720.09(1)** | **−9720.09** | **−9717.96 ± 0.99** | **251.6 ± 28.9** |
| **0.12** | **−9717.46(6)** | **−9720.09(1)** | **−9720.09** | **−9717.97 ± 0.80** | 261.4 ± 21.2 |
| **0.14** | **−9717.46(8)** | **−9720.09(1)** | **−9720.09** | **−9717.84 ± 4.71** | 273.4 ± 31.5 |
| **0.15** | **−9717.46(9)** | **−9720.09(1)** | **−9720.09** | **−9717.73 ± 0.79** | 262.6 ± 36.3 |
| **0.18** | **−9717.46(9)** | −9717.46(9) | −9725.60 | −9718.28 ± 2.44 | 271.7 ± 52.6 |
| 0.20 | −9717.46(4) | −9720.09(2) | −9731.03 | −9720.54 ± 4.32 | 280.6 ± 34.8 |
| 0.22 | −9717.46(5) | −9720.09(2) | −9721.28 | −9721.28 ± 1.03 | 289.3 ± 28.9 |
| 0.25 | −9717.46(5) | −9720.09(1) | −9725.53 | −9719.97 ± 2.46 | 303.3 ± 30.7 |

a The numbers in parentheses indicate the number to obtain the corresponding K2 scores.
b Results in the form μ±δ indicate the mean $\mu$ and the standard deviation $\delta$.

**Table 8.** Significance ($p$-values [a] from Kruskal–Wallis tests) for $P_{ed}$.

| $P_{ed}$ | 0.10 | 0.12 | 0.14 | 0.15 | 0.18 |
|---|---|---|---|---|---|
| 0.10 |  | 0.7976 | 0.3746 | 0.1632 | 0.1962 |
| 0.12 |  |  | 0.4280 | 0.1796 | 0.1966 |
| 0.14 |  |  |  | 0.5839 | 0.6267 |
| 0.15 |  |  |  |  | 0.9422 |
| 0.18 |  |  |  |  |  |

a The confidence level is 95%.

ered popular networks of medium size: Alarm, Insurance, and Child. Thus, we selected three datasets generated from Alarm, Insurance, and Child: Alarm-2000, Insurance-3000, and Child-2000, respectively. Chemotaxis is a critical driving force for the BFO algorithm, without which BFO cannot work properly. Therefore, we used the chemotaxis mechanism as the essential mechanism of the algorithm and explored the individual contributions of the other four operators. Specifically, the four variations are

BFO-B1: remove the reproduction process from BFO-B.

BFO-B2: remove the elimination and dispersal process from BFO-B.

BFO-B3: remove the move operator from BFO-B.

BFO-B4: remove the swimming process from BFO-B.

BFO-B1 and BFO-B2 test the effects of the reproduction and elimination and dispersal mechanisms, respectively. BFO-B3 validates the effectiveness of the move operator, which is a non-standard operator, whereas the other three operators, addition, deletion, and reversion, are standard and essential for the score and search approach when learning BN structures. Swimming means repeatedly optimizing a solution using the same operator. By removing the swimming process, BFO-B4 optimizes a solution with a random mix of four different operators, which may be used to explore the effect of the swimming process. For each dataset, 20 independent trials were executed, as summarized in Table 9, where the definitions of the other items are as in Section 4.2, and the new evaluation metrics are

SSD: the Smallest Structural Difference over all trials, i.e., the smallest number of arcs wrongly added, deleted and reversed over all trials. The smaller the corresponding value, the better the learned network.

**Table 9.** Performance of the BFO-B algorithm and variations.

| Dataset | Algorithm | Scoring | | | Structure | | |
|---|---|---|---|---|---|---|---|
| | | HKS | LKS | AKS[b] | SSD | BSD | ASD[b] |
| Alarm-2000 | BFO-B[a] | −9717.46 | −9720.09 | **−9717.95 ± 1.08** | 3 | 7 | **5.05 ± 1.20** |
| | BFO-B1 | −9717.46 | −9723.34 | −9718.86 ± 1.78 | 3 | 10 | 5.25 ± 2.12 |
| | BFO-B2 | −9717.46 | −9725.97 | −9718.72 ± 2.67 | 3 | 11 | 5.60 ± 2.03 |
| | BFO-B3 | −9717.46 | −9729.62 | −9720.25 ± 3.42 | 3 | 13 | 7.75 ± 3.99 |
| | BFO-B4 | −9717.46 | −9734.95 | −9719.41 ± 4.44 | 3 | 14 | 6.70 ± 2.20 |
| Insurance-3000 | BFO-B[a] | **−19 862.83** | −19 930.66 | **−19 887.29 ± 22.93** | 2 | **12** | **7.15 ± 3.00** |
| | BFO-B1 | −19 865.08 | **−19 930.13** | −19 895.55 ± 17.99 | 3 | 13 | 7.25 ± 2.12 |
| | BFO-B2 | −19 864.74 | −20 071.83 | −19 902.72 ± 44.87 | 3 | 17 | 8.05 ± 8.31 |
| | BFO-B3 | −19 865.08 | −20 080.80 | −19 918.56 ± 65.07 | 3 | 19 | 9.35 ± 4.69 |
| | BFO-B4 | −19 867.96 | −20 063.56 | −19 925.63 ± 56.00 | 3 | 23 | 9.50 ± 5.57 |
| Child-2000 | BFO-B[a] | **−10 715.67** | **−10 716.37** | **−10 715.77 ± 0.25** | 1 | 4 | **1.45 ± 1.07** |
| | BFO-B1 | −10 715.67 | −10 716.37 | −10 715.81 ± 0.29 | 1 | 4 | 1.60 ± 1.23 |
| | BFO-B2 | −10 715.67 | −10 718.61 | −10 716.06 ± 0.69 | 1 | 4 | 2.10 ± 1.45 |
| | BFO-B3 | −10 715.67 | −10 716.66 | −10 716.07 ± 0.38 | 1 | 4 | 2.60 ± 1.50 |
| | BFO-B4 | −10 715.67 | −10 718.61 | −10 715.84 ± 0.69 | 1 | 4 | 1.75 ± 1.23 |

a BFO-B parameter values were $S=80$, $N_s=4$, $N_c=30$, $N_{re}=4$, $N_{ed}=3$, and $P_{ed}=0.10$.
b Results in the form $\mu \pm \delta$ indicate the mean $\mu$ and the standard deviation $\delta$.

BSD: the Biggest Structural Difference over all trials.

ASD: the Average overall Structural Difference over all trials.

Both AKS and ASD show that the BFO-B variants achieve significantly inferior results than the whole BFO-B algorithm on all three datasets. Thus, the corresponding four components are all effective in assisting the BFO-B algorithm to learn a better solution. The reasons for their effectiveness can be concluded as follows: (1) Reproduction takes charge of information transmission among individuals, which is a key factor for a swarm intelligence method to obtain a better solution. (2) Elimination and dispersal makes some individuals search for a better solution in new regions, which plays a main role in global search, and is important for the algorithm to maintain a balance between global exploration and local exploitation. (3) Compared with the three standard operators (addition, deletion, and reversion), move makes a bigger change on a candidate solution, which offers more chances to assist the algorithm in escaping from a local optima. (4) Under a certain number of chemotaxis steps, the swimming process makes it possible to perform more explorations around a candidate solution and thereby is useful for the algorithm. From Table 9, we also observe that BFO-B1 and BFO-B2 have less difference from the whole BFO-B algorithm for AKS and ASD, which indicates that although reproduction and elimination and dispersal have positive roles in enhancing the performance of the algorithm, chemotaxis is the most essential component for the BFO-B algorithm. BFO-B3 and BFO-B4 show larger differences from the whole BFO-B algorithm. Since the move operator and swimming process are two key strategies in chemotaxis, removing either of them is equivalent to damaging the chemotaxis process. Hence the size of these differences provides further evidence of the importance of chemotaxis to the proposed BFO-B algorithm.

Thus, we conclude that reproduction, elimination and dispersal, move, and swimming processes are essential components for BFO-B to efficiently learn a BN structure from data.

## 4.4. Learning BNs using BFO-B

To fully test the performance of the proposed BFO-B algorithm, we ran 20 independent trials on all the datasets in Table 2, as summarized in Table 10 for K2 score and time performance, and Table 11 for structural differ-

Table 10. K2 score and time performance of BFO-B[a] on different datasets.

| Datasets | HKS | LKS | KLQ | AKS[b] | AET[b] |
|---|---|---|---|---|---|
| Alarm-1000 | −5026.34 | −5031.72 | −5029.18 | −5026.44 ± 2.38 | 191.8 ± 37.8 |
| Alarm-2000 | −9717.46 | −9720.09 | −9717.46 | −9717.95 ± 1.08 | 246.9 ± 24.6 |
| Alarm-3000 | −14401.28 | −14409.31 | −14404.38 | −11403.58 ± 1.75 | 339.0 ± 43.0 |
| Alarm-4000 | −19098.59 | −19108.94 | −19101.94 | −19099.78 ± 2.55 | 429.2 ± 37.8 |
| Alarm-5000 | −23782.15 | −23790.42 | −23782.15 | −23783.28 ± 2.21 | 537.4 ± 35.4 |
| Alarm-6000 | −28347.11 | −28350.14 | −28347.21 | −28347.65 ± 0.91 | 631.1 ± 38.8 |
| Alarm-7000 | −33022.93 | −33025.13 | −33023.09 | −33023.39 ± 0.73 | 726.9 ± 44.7 |
| Alarm-8000 | −37745.54 | −37762.08 | −37747.40 | −37747.31 ± 3.72 | 832.2 ± 53.3 |
| Insurance-3000 | −19862.83 | −19930.66 | −19891.27 | −19887.29 ± 22.93 | 341.8 ± 20.9 |
| Insurance-5000 | −32372.60 | −32468.59 | −32426.84 | −32410.79 ± 26.42 | 793.2 ± 77.2 |
| Insurance-6000 | −38580.79 | −38790.73 | −38647.09 | −38635.78 ± 51.45 | 996.2 ± 62.2 |
| Child-2000 | −10715.67 | −10716.37 | −10715.67 | −10715.77 ± 0.25 | 35.9 ± 4.9 |
| Child-5000 | −26574.91 | −26575.58 | −26574.91 | −26574.98 ± 0.20 | 88.2 ± 7.8 |
| Credit-3000 | −13842.16 | −13844.84 | −13842.16 | −13842.48 ± 0.79 | 10.1 ± 12.4 |
| Credit-6000 | −27546.22 | −27554.97 | −27550.88 | −27550.58 ± 1.95 | 17.1 ± 11.8 |
| Asia-1000 | −1011.26 | −1011.76 | −1011.26 | −1011.38 ± 0.22 | 0.6 ± 1.2 |
| Asia-5000 | −4883.39 | −4883.39 | −4883.39 | −4883.39 ± 0.00 | 1.8 ± 0.4 |
| Child3-5000 | −81754.96 | −81760.36 | −81756.75 | −81756.39 ± 1.91 | 1368.0 ± 62.5 |
| Alarm3-5000 | −75340.77 | −75355.96 | −75348.57 | −75344.67 ± 5.53 | 8530.1 ± 501.1 |
| EngineFuelSystem-10000 | −2271.71 | −2271.71 | −2271.71 | −2271.71 ± 0.00 | 3.0 ± 0.0 |
| Boerlage92-10000 | −44205.85 | −44208.68 | −44206.33 | −44206.42 ± 0.99 | 106.00 ± 3.64 |
| Studfarm-10000 | −1662.17 | −1662.17 | −1662.17 | −1662.17 ± 0.00 | 25.9 ± 1.9 |
| Brain-10000 | −290705.58 | −290742.92 | −290709.58 | −290714.57 ± 11.42 | 259.0 ± 34.4 |
| Tank-100000 | −14251.09 | −14252.30 | −14252.30 | −14252.15 ± 0.37 | 136.4 ± 3.1 |
| Win95pts-50000 | −196805.89 | −196807.39 | −196812.38 | −196806.89 ± 1.49 | 58768.5 ± 1634.6 |
| Hepar II-50000 | −705521.93 | −705539.43 | −705533.47 | −705528.31 ± 6.50 | 5925.7 ± 146.9 |

a BFO-B parameters values were S=80, $N_s$=4, $N_c$=30, $N_{re}$=4, $N_{ed}$=3, and $P_{ed}$=0.10.
b Results in the form $\mu \pm \delta$ indicate the mean $\mu$ and the standard deviation $\delta$.

ence between the learned and the true networks. The new evaluation metrics used in Table 10 and Table 11 are defined below, and the meanings of the other items are as in Sections 4.2 and 4.3.

KLQ: the Lower Quartile of the K2 scoring over all trials.

SLQ: the Lower Quartile of overall Structural difference over all trials sorted.

AAD: the Average Difference of arcs incorrectly Added over all trials, i.e., the average number of the arcs wrongly added over all trials.

ADD: the Average Difference of arcs incorrectly Deleted over all trials.

ARD: the Average Difference of arcs incorrectly Reversed over all trials.

The difference between HKS and LKS is small (Table 10), which indicates that the BFO-B algorithm is stable for all of the datasets. Also, KLQ is equal to the corresponding HKS for Alarm-2000, Alarm-5000, Child-2000, Child-5000, Credit-3000, Asia-1000, Asia-5000, EngineFuelSystem-10000, and Studfarm-10000, which means that the proposed BFO-B algorithm returns the best networks on at least 15 of the 20 runs on these datasets. HKS and LKS are the same for Asia-5000, EngineFuelSystem-10000, and Studfarm-10000, which means that BFO-B obtains exactly the same results over 20 runs. Thus, the proposed BFO-B algorithm performs well for relatively smaller networks. HKS, KLQ, and AKS for relatively large networks (Child3-5000, Alarm3-5000, Brain-10000, Win95pts-50000, and Hepar II-50000) are higher than for the corresponding true networks (Table 2), which shows that even for larger networks, the BFO-B algorithm is capable of finding networks with higher K2 scores (i.e., the networks which most match the datasets), provided they exist in the search space.

Mean and standard deviations of AAD, ADD, ARD and ASD (Table 11) are relatively small for datasets generated from the nine smaller networks (Alarm, Insurance, Child, Credit, Asia, EngineFuelSystem, Boeralge92, Studfarm, and Tank). The best values (numbers in parentheses) are not more than 3, except for Alarm-1000, which does not contain enough cases to correctly learn a BN structure. It is interesting that all of the statistics on Studfarm-10000 are 0, indicating that BFO-B always found the true network over 20 runs. Thus, the proposed BFO-B algorithm finds networks with structures very similar to the true structures for smaller networks. For datasets Child3-5000, Alarm3-5000, Brain-10000, Win95pts-50000, and

**Table 11.** Structural difference performance of BFO-B[a] on different datasets.

| Datasets | AAD[b] | ADD[b] | ARD[b] | ASD[b] | SLQ |
|---|---|---|---|---|---|
| Alarm-1000 | 7.50 ± 2.66 (4) | 2.05 ± 0.92 (1) | 4.80 ± 2.86 (1) | 14.35 ± 5.29 (6) | 17 |
| Alarm-2000 | 2.90 ± 0.44 (2) | 1.00 ± 0.00 (1) | 1.15 ± 0.85 (0) | 5.05 ± 1.20 (3) | 5 |
| Alarm-3000 | 0.80 ± 0.68 (0) | 1.10 ± 0.30 (1) | 1.05 ± 1.20 (0) | 2.85 ± 1.98 (1) | 3 |
| Alarm-4000 | 2.05 ± 0.86 (1) | 1.00 ± 0.00 (1) | 1.15 ± 1.24 (0) | 4.20 ± 2.14 (2) | 4 |
| Alarm-5000 | 1.05 ± 0.92 (0) | 1.00 ± 0.00 (1) | 1.15 ± 1.19 (0) | 3.20 ± 2.17 (1) | 3 |
| Alarm-6000 | 1.90 ± 0.44 (0) | 1.00 ± 0.00 (1) | 0.95 ± 0.50 (0) | 2.85 ± 0.91 (1) | 3 |
| Alarm-7000 | 0.90 ± 0.44 (0) | 1.00 ± 0.00 (1) | 0.90 ± 0.44 (0) | 2.80 ± 0.87 (1) | 3 |
| Alarm-8000 | 1.20 ± 0.75 (0) | 1.00 ± 0.00 (1) | 1.45 ± 1.20 (0) | 3.55 ± 2.01 (1) | 3 |
| Insurance-3000 | 3.10 ± 1.29 (1) | 0.25 ± 0.44 (0) | 3.80 ± 1.67 (1) | 7.15 ± 3.00 (2) | 9 |
| Insurance-5000 | 2.30 ± 1.26 (1) | 0.20 ± 0.41 (0) | 3.60 ± 2.23 (1) | 6.10 ± 3.26 (2) | 7 |
| Insurance-6000 | 2.95 ± 1.73 (1) | 0.20 ± 0.41 (0) | 4.20 ± 2.31 (1) | 7.35 ± 4.11 (2) | 9 |
| Child-2000 | 0.00 ± 0.00 (0) | 0.00 ± 0.00 (0) | 1.45 ± 1.07 (1) | 1.45 ± 1.07 (1) | 1 |
| Child-5000 | 0.00 ± 0.00 (0) | 0.00 ± 0.00 (0) | 1.30 ± 0.90 (1) | 1.30 ± 0.90 (1) | 1 |
| Credit-3000 | 0.95 ± 0.39 (0) | 0.00 ± 0.00 (0) | 2.85 ± 0.67 (1) | 3.90 ± 0.79 (1) | 4 |
| Credit-6000 | 1.05 ± 0.59 (0) | 0.00 ± 0.00 (0) | 2.10 ± 1.18 (0) | 3.15 ± 1.81 (0) | 3 |
| Asia-1000 | 1.25 ± 0.43 (1) | 1.00 ± 0.00 (1) | 1.15 ± 0.36 (1) | 3.40 ± 0.73 (3) | 3 |
| Asia-5000 | 0.00 ± 0.00 (0) | 0.00 ± 0.00 (0) | 1.00 ± 0.00 (1) | 1.00 ± 0.00 (1) | 1 |
| Child3-5000 | 2.05 ± 0.22 (2) | 0.35 ± 0.67 (0) | 4.55 ± 2.16 (3) | 6.95 ± 2.67 (5) | 7 |
| Alarm3-5000 | 6.55 ± 1.00 (5) | 3.15 ± 0.88 (2) | 6.45 ± 2.39 (3) | 16.15 ± 2.80 (13) | 18 |
| EngineFuelSystem-10000 | 0.00 ± 0.00 (0) | 0.00 ± 0.00 (0) | 2.00 ± 0.00 (0) | 2.00 ± 0.00 (0) | 2 |
| Boerlage92-10000 | 0.45 ± 0.51 (0) | 6.90 ± 0.94 (6) | 5.30 ± 0.47 (5) | 12.65 ± 1.31 (11) | 14 |
| Studfarm-10000 | 0.00 ± 0.00 (0) | 0.00 ± 0.00 (0) | 0.00 ± 0.00 (0) | 0.00 ± 0.00 (0) | 0 |
| Brain-10000 | 0.00 ± 0.00 (0) | 0.00 ± 0.00 (0) | 17.05 ± 2.96 (14) | 17.05 ± 2.96 (14) | 19 |
| Tank-10000 | 1.00 ± 0.00 (0) | 0.00 ± 0.00 (0) | 2.20 ± 0.60 (2) | 3.20 ± 1.40 (2) | 4 |
| Win95pts-50000 | 22.5 ± 2.52 (20) | 14.85 ± 0.81 (14) | 8.45 ± 1.50 (7) | 45.80 ± 4.77 (41) | 48 |
| Hepar II-50000 | 6.20 ± 1.44 (5) | 11.70 ± 0.57 (10) | 10.75 ± 2.15 (9) | 28.65 ± 3.34 (26) | 31 |

a BFO-B parameters values were $S=80$, $N_s=4$, $N_c=30$, $N_{re}=4$, $N_{ed}=3$, and $P_{ed}=0.10$.

b Results in the form $\mu \pm \delta$ indicate the mean $\mu$ and the standard deviation $\delta$; the numbers in parentheses indicate the best values obtained.

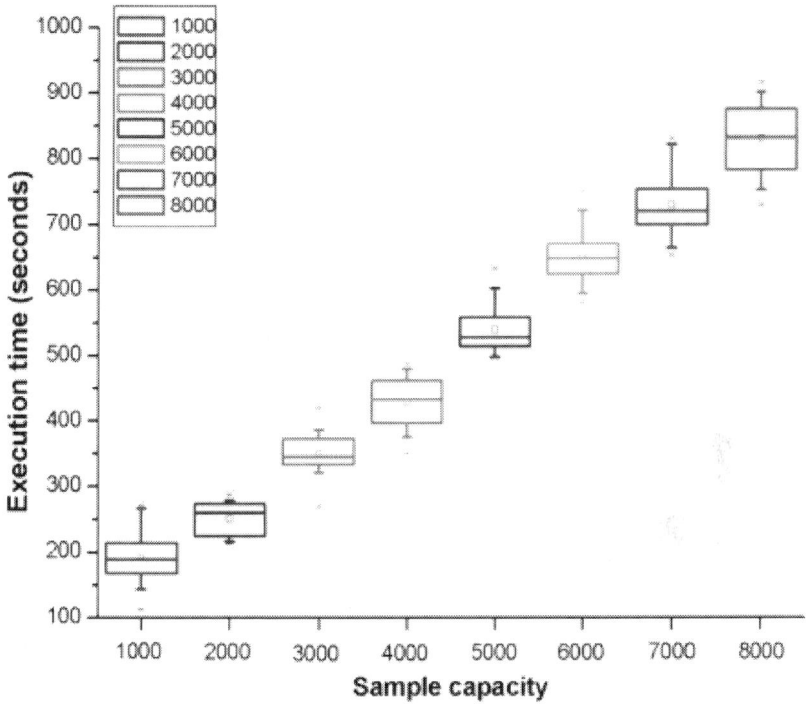

**Figure 3.** BFO-B time performance on the Alarm network.

Hepar II-50000, generated from larger networks, the BFO-B algorithm does not perform well, but still attains relatively good results on Child3-5000, Alarm3-5000 and Brain-10000. For Win95pts-50000 and Hepar II-50000, the BFO-B algorithm obtains networks with more incorrect arcs compared to the true networks, because the Win95pts and Hepar II networks have much more complex network structures (some nodes have more parent nodes or child nodes), and 50000 available cases does not fully reflect their network characteristics. From the Alarm, Insurance, Child, and Credit datasets, BFO-B outcomes improve as data volume increases. This illustrates that larger data volumes are more conducive to correctly learning the underlying network structures. The tiled networks, Alarm and Child, show that BFO-B performance decreases as the number of variables increases while the difficulty of learning the network remains constant.

Execution times for the BFO-B algorithm on different datasets generated from the Alarm network is shown in Fig. 3 in the form of a box plot, where the top and the bottom of each box indicate the 75th and 25th percentiles, respectively; the line in each box indicates the 50th percentile; the whisker bars below and above each box indicate the 10th and 90th percentiles, respectively; and the squares and asterisks in each box indicate the mean and outliers, respectively. The mean values are all close to 50th percentile for all datasets except Alarm-2000 and Alarm-5000, and execution time fluctuates weakly on each dataset, which indicates that the BFO-B algorithm runs stably. Execution time grows relatively slowly with increasing sample capacity, suggesting that the BFO-B algorithm is capable of handling relatively large datasets.

Considering K2 scores, structural differences and execution time, we conclude that the proposed BFO-B algorithm can find network structures with higher K2 scores and smaller structural differences, and relatively stable runtime efficiency. Thus, the proposed BFO-B algorithm appears to be a more promising algorithm for learning BNs in the field of big data.

## 4.5. Comparing BFO-B with other Algorithms

We compared the proposed BFO-B algorithm with the nine well-known algorithms: ABC-B [10], ACO-B [6], GA-B [9], HCST-B [26], SC [21], GES [28], PC [29], TPDA [19] and MMHC [15]. These algorithms incorporate almost all the different classes of methods for learning BN structures from data and are all well-known methods in the domain of BN structural learning:

ACO-B and ABC-B are state of the art swarm intelligence based methods recently developed that simulate foraging behavior of real ants and honeybees, respectively.

GA-B is a classic evolution based algorithm, which makes use of a genetic algorithm to search for node ordering of the system variables and then evaluates the quality of the variable ordering with the K2 algorithm.

HCST-B is a simple and common heuristic based algorithm that uses a hill climbing algorithm to construct network structures.

SC constrains the search space by restricting the parents of each variable to belong to a small subset of candidates and also makes use of a hill climbing algorithm to learn network structures.

GES provides a new implementation of the search space, using equivalence classes as states in a greedy search.

PC is a prototypical dependency analysis algorithm.

TPDA employs the concept of mutual information to test for conditional independencies.

MMHC is a hybrid method combing dependency analysis and score and search, which uses conditional independence tests to identify the skeleton of the network structure and then orients the skeleton using a greedy hill climbing search.

The first six algorithms (ACO-B, ABC-B, GA-B, HCST-B, SC, and GES) are score and search approaches, PC and TPDA are dependency analysis approaches, and MMHC is a hybrid approach. Because of space limitations, we report eight representative datasets generated from different benchmark networks with different sizes: Alarm-6000, Insurance-3000, Child-2000, Asia-5000, Studfarm-10000, Boerlage92-1000, Brain-10000, and Hepar II-50000.

To ensure fair comparison, the population sizes of the three population based algorithms (ABC-B, ACO-B, and GA-B) were set to 80. The other specific parameters of these algorithms and the HCST-B algorithm conformed to the best settings as reported in their original papers, which were tuned by a series of experiments. We obtained software implementations of the SC, PC, TPDA, and MMHC algorithms in the Causal Explorer system,[1] and software implementation of the GES algorithm from the TETRAD Project.[2] We used the default values in the software implementations for the parameters of these algorithms. Each algorithm was executed for 20 independent runs over each dataset, as summarized in Table 12 and Table 13 for K2 score and time aspects, along with the detailed parameter values for some of the algorithms. Since PC and TPDA are based on independency analysis, which do not use the scoring mechanism in the learning process, but to uniformly compare the different algorithms, we computed the scores of the final networks learned by them. Table 14 and Table 15 show the structural difference outcomes, where the numbers in parentheses represent the best result over 20 runs. The meanings of all of the metrics in Table 12, Table 13, Table 14 and Table 15 are as described variously above, and the best value for each metric except AET is shown in bold. We also performed pair-wise Kruskal–Wallis tests for the BFO-B algorithm against each of the other nine algorithms on each dataset to determine significant differences, as shown in Fig. 4(a) and 4(b) for K2 scores and overall structural difference, respectively. The red line is the benchmark, $p=0.05$. When the re-

**Table 12.** K2 score and time comparisons among 10 algorithms on Alarm-6000, Insurance-3000, Child-2000, and Asia-5000.

| Datasets | Algorithms | HKS | LKS | KLQ | AKS[b] | AET[b] (s) |
|---|---|---|---|---|---|---|
| Alarm-6000 | BFO-B[a] | −28 347.11 | **−28 350.14** | −28 347.21 | −28 347.65 ± 0.91 | 631.1 ± 38.8 |
| | ABC-B[b] | −28 347.11 | −28 366.37 | −28 348.27 | −28 348.14 ± 4.18 | 266.5 ± 21.4 |
| | ACO-B[c] | −28 347.11 | −28 369.51 | −28 347.38 | −28 348.56 ± 4.84 | 810.1 ± 120.4 |
| | GA-B[d] | −28 368.77 | −28 431.92 | −28 389.06 | −28 394.30 ± 18.05 | 3068.6 ± 118.3 |
| | HCST-B[e] | −28 710.68 | −28 710.68 | −28 710.682 | −28 710.68 ± 0.00 | 14.2 ± 0.4 |
| | SC[f] | −35 748.22 | −35 748.22 | −35 748.22 | −35 748.22 ± 0.00 | 71.8 ± 0.1 |
| | GES[g] | −28 923.47 | −28 923.47 | −28 923.47 | −28 923.47 ± 0.00 | 4.6 ± 0.1 |
| | PC[g] | −37 091.14 | −37 091.14 | −37 091.14 | −37 091.14 ± 0.00 | 5.6 ± 0.06 |
| | TPDA[f] | −33 138.98 | −33 138.98 | −33 138.98 | −33 138.98 ± 0.00 | 5.8 ± 0.04 |
| | MMHC[f] | −28 497.40 | −28 497.40 | −28 497.40 | −28 497.40 ± 0.00 | 3.2 ± 0.8 |
| Insurance-3000 | BFO-B | **−19 862.83** | −19 930.66 | −19 891.27 | −19 887.29 ± 22.93 | 341.8 ± 20.9 |
| | ABC-B | −19 865.08 | **−19 921.65** | −19 901.98 | −19 895.29 ± 20.55 | 135.3 ± 8.9 |
| | ACO-B | **−19 862.83** | −19 933.90 | −19 907.90 | −19 894.40 ± 22.90 | 332.8 ± 64.31 |
| | GA-B | −19 864.84 | −20 218.20 | −19 949.89 | −19 925.39 ± 83.03 | 1197.4 ± 19.22 |
| | HCST-B | −21 840.47 | −21 840.47 | −21 840.47 | −21 840.47 ± 0.00 | 5.2 ± 1.1 |
| | SC | −26 885.15 | −26 885.15 | −26 885.15 | −26 885.15 ± 0.00 | 8.1 ± 0.1 |
| | GES | −20 395.00 | −20 395.00 | −20 395.00 | −20 395.00 ± 0.00 | 3.9 ± 0.2 |
| | PC | −27 088.96 | −27 088.96 | −27 088.96 | −27 088.96 ± 0.00 | 4.8 ± 0.06 |
| | TPDA | −26 990.76 | −26 990.76 | −26 990.76 | −26 990.76 ± 0.00 | 4.4 ± 0.02 |
| | MMHC | −22 024.68 | −22 024.68 | −22 024.68 | −22 024.68 ± 0.00 | 16.5 ± 0.6 |

| | | | | | |
|---|---|---|---|---|---|
| Child-2000 | BFO-B | −10715.67 | −10716.37 | −10715.77 ± 0.25 | 26.9 ± 4.2 |
| | ABC-B | **−10715.67** | −10716.37 | −10715.84 ± 0.31 | 7.4 ± 2.4 |
| | ACO-B | **−10715.67** | **−10715.67** | **−10715.67 ± 0.00** | 10.8 ± 3.0 |
| | GA-B | **−10715.67** | **−10715.67** | **−10715.67 ± 0.00** | 105.4 ± 2.0 |
| | HCST-B | −10716.66 | −10716.66 | −10716.66 ± 0.00 | 1.3 ± 0.1 |
| | SC | −11340.89 | −11340.89 | −11340.89 ± 0.00 | 5.1 ± 0.1 |
| | GES | −11585.60 | −11585.60 | −11585.60 ± 0.00 | 2.5 ± 0.05 |
| | PC | −12649.87 | −12649.87 | −12649.87 ± 0.00 | 3.1 ± 0.02 |
| | TPDA | −12345.64 | −12345.64 | −12345.64 ± 0.00 | 2.1 ± 0.01 |
| | MMHC | −10733.11 | −10733.11 | −10733.11 ± 0.00 | 1.4 ± 0.3 |
| Asia-5000 | BFO-B | **−4883.39** | **−4883.39** | **−4883.39 ± 0.00** | 1.8 ± 0.4 |
| | ABC-B | **−4883.39** | **−4883.39** | **−4883.39 ± 0.00** | 0.5 ± 0.04 |
| | ACO-B | **−4883.39** | **−4883.39** | **−4883.39 ± 0.00** | 1.0 ± 0.0 |
| | GA-B | **−4883.39** | −4884.66 | −4884.08 ± 0.22 | 4.1 ± 0.1 |
| | HCST-B | −4885.71 | −4885.71 | −4885.71 ± 0.00 | 0.5 ± 0.1 |
| | SC | −4883.75 | −4883.75 | −4883.75 ± 0.00 | 0.9 ± 0.04 |
| | GES | −4884.05 | −4884.05 | −4884.05 ± 0.00 | 0.9 ± 0.03 |
| | PC | −5533.11 | −5533.11 | −5533.11 ± 0.00 | 1.8 ± 0.01 |
| | TPDA | −5160.60 | −5160.60 | −5160.60 ± 0.00 | 1.0 ± 0.01 |
| | MMHC | −4883.44 | −4883.44 | −4883.44 ± 0.00 | 0.2 ± 0.06 |

a BFO-B parameters were $S=80$, $N_s=4$, $N_c=30$, $N_{re}=4$, $N_{cd}=3$ and $P_{ed}=0.10$.
b ABC-B parameters were $K=80$, $\alpha=1$, $\beta=2$, $\rho=0.1$, $q_0=0.8$, $N=150$ and $limit=3$.
c ACO-B parameters were $m=80$, $\rho=\psi=0.4$, $\beta=2.0$, $q_0=0.8$ and $tmax=150$.
d GA-B parameters were $\lambda=80$, $p_c=1$ and $p_m=0.01$.
e HCST-B parameter was $maxiter=1200$.
f Using the default parameter values for the corresponding algorithms in the Causal Explorer system.
g Using the default parameter values for the GES algorithm in the TETRAD Project.
h Results in the form $\mu \pm \delta$ indicate the mean $\mu$ and the standard deviation $\delta$.

**Table 13.** K2 score and time comparisons among 10 algorithms on Studfarm-10000, Boerlage92-10000, Brain-10000, and Hepar II-50000.

| Datasets | Algorithms | HKS | LKS | KLQ | AKS | AET (s) |
|---|---|---|---|---|---|---|
| Studfarm-10000 | BFO-B | **−1662.17** | **−1662.17** | **−1662.17** | **−1662.17 ± 0.00** | 25.9 ± 1.9 |
| | ABC-B | **−1662.17** | **−1662.17** | **−1662.17** | **−1662.17 ± 0.00** | 9.3 ± 1.2 |
| | ACO-B | **−1662.17** | **−1662.17** | **−1662.17** | **−1662.17 ± 0.00** | 21.1 ± 1.7 |
| | GA-B | **−1662.17** | −1663.07 | −1662.81 | −1662.69 ± 0.28 | 23.9 ± 0.4 |
| | HCST-B | −1780.26 | −1780.26 | −1780.26 | −1780.26 ± 0.00 | 0.8 ± 0.04 |
| | SC | −1935.42 | −1935.42 | −1935.42 | −1935.42 ± 0.00 | 2.1 ± 0.04 |
| | GES | −1808.84 | −1808.84 | −1808.84 | −1808.84 ± 0.00 | 1.6 ± 0.3 |
| | PC | −2155.22 | −2155.22 | −2155.22 | −2155.22 ± 0.00 | 0.8 ± 0.02 |
| | TPDA | −1924.33 | −1924.33 | −1924.33 | −1924.33 ± 0.00 | 1.0 ± 0.02 |
| | MMHC | −1662.42 | −1662.42 | −1662.42 | −1662.42 ± 0.00 | 1.4 ± 0.6 |
| Boerlage92-10000 | BFO-B | **−44205.85** | −44208.68 | −44206.33 | **−44206.32 ± 0.99** | 106.00 ± 3.64 |
| | ABC-B | **−44205.85** | −44208.68 | −44208.68 | −44206.73 ± 1.32 | 35.1 ± 6.4 |
| | ACO-B | **−44205.85** | −44208.85 | **−44205.85** | −44206.42 ± 1.09 | 78.0 ± 4.9 |
| | GA-B | −44208.07 | −44220.79 | −44214.68 | −44212.04 ± 3.47 | 162.8 ± 5.4 |
| | HCST-B | −44223.11 | −44223.11 | −44223.11 | −44223.11 ± 0.00 | 3.5 ± 0.2 |
| | SC | −50468.37 | −50468.37 | −50468.37 | −50468.37 ± 0.00 | 7.1 ± 0.5 |
| | GES | −44817.10 | −44817.10 | −44817.10 | −44817.10 ± 0.00 | 3.1 ± 0.1 |
| | PC | −46462.16 | −46462.16 | −46462.16 | −46462.16 ± 0.00 | 3.4 ± 0.6 |
| | TPDA | −46117.33 | −46117.33 | −46117.33 | −46117.33 ± 0.00 | 1.8 ± 0.2 |
| | MMHC | −44215.80 | −44215.80 | −44215.80 | −44215.80 ± 0.00 | 2.6 ± 0.4 |

**Table 13.** (Continued)

| | | | | | |
|---|---|---|---|---|---|
| Brain-10000 | BFO-B | −290 709.58 | −290 742.92 | −290 709.58 | −290 714.57 ± 11.42 | 259.0 ± 34.4 |
| | ABC-B | −290 709.58 | −290 770.44 | −290 729.16 | −290 719.03 ± 15.21 | 188.2 ± 18.9 |
| | ACO-B | −290 709.58 | −290 736.34 | −290 709.58 | **−290 711.82 ± 6.48** | 259.0 ± 34.4 |
| | GA-B | −290 864.22 | −290 998.45 | −290 920.35 | −290 869.85 ± 52.48 | 211.1 ± 1.9 |
| | HCST-B | −291 080.26 | −291 080.26 | −291 080.26 | −291 080.26 ± 0.00 | 16.5 ± 0.2 |
| | SC | −292 297.57 | −292 297.57 | −292 297.57 | −292 297.57 ± 0.00 | 178.5 ± 4.8 |
| | GES | −292 541.46 | −292 541.46 | −292 541.46 | −292 541.46 ± 0.00 | 9.6 ± 1.1 |
| | PC | −301 049.56 | −301 049.56 | −301 049.56 | −301 049.56 ± 0.00 | 28 813.7 ± 1420.4 |
| | TPDA | −296 390.72 | −296 390.72 | −296 390.72 | −296 390.72 ± 0.00 | 238.6 ± 3.3 |
| | MMHC | −296 442.64 | −296 442.64 | −296 442.64 | −296 442.64 ± 0.00 | 91 141.5 ± 1200.5 |
| Hepar II-50000 | BFO-B | −705 521.93 | −705 539.43 | −705 533.47 | −705 528.31 ± 6.50 | 5925.7 ± 146.9 |
| | ABC-B | **−705 517.65** | −705 536.82 | **−705 529.37** | **−705 528.03 ± 4.67** | 5605.8 ± 707.2 |
| | ACO-B | −705 521.93 | **−705 535.74** | −705 530.08 | −705 528.79 ± 4.38 | 11 655.4 ± 1763.4 |
| | GA-B | −705 540.11 | −705 552.82 | −705 550.40 | −705 552.82 ± 4.91 | 38 146.6 ± 1247.1 |
| | HCST-B | −706 471.14 | −706 471.14 | −706 471.14 | −706 471.14 ± 0.00 | 189.9 ± 1.42 |
| | SC | −752 658.41 | −752 658.41 | −752 658.41 | −752 658.41 ± 0.00 | 739.7 ± 15.4 |
| | GES | −711 162.49 | −711 162.49 | −711 162.49 | −711 162.49 ± 0.00 | 98.4 ± 5.6 |
| | PC | −731 214.76 | −731 214.76 | −731 214.76 | −731 214.76 ± 0.00 | 14.7 ± 1.2 |
| | TPDA | −730 291.66 | −730 291.66 | −730 291.66 | −730 291.66 ± 0.00 | 9.6 ± 0.2 |
| | MMHC | −706 214.62 | −706 214.62 | −706 214.62 | −706 214.62 ± 0.00 | 6750.3 ± 124.3 |

**Table 14.** Structural differences among 10 algorithms on Alarm-6000, Insurance-3000, Child-2000, and Asia-5000.

| Datasets | Algorithms | AAD | ADD | ARD | ASD | SLQ |
|---|---|---|---|---|---|---|
| Alarm-6000 | BFO-B | 0.90 ± 0.44 (0) | 1.00 ± 0.00 (1) | 0.95 ± 0.50 (0) | 2.85 ± 0.91 (1) | 3 |
| | ABC-B | 1.75 ± 0.54 (1) | 1.05 ± 0.22 (1) | 1.80 ± 0.68 (1) | 4.60 ± 1.36 (3) | 5 |
| | ACO-B | 1.60 ± 0.50 (1) | 1.05 ± 0.22 (1) | 1.95 ± 1.12 (1) | 4.60 ± 1.62 (3) | 5 |
| | GA-B | 3.40 ± 1.79 (2) | 1.95 ± 0.39 (1) | 4.45 ± 1.88 (3) | 9.80 ± 3.49 (7) | 10 |
| | HCST-B | 4.00 ± 0.00 (4) | 4.00 ± 0.00 (4) | 6.00 ± 0.00 (6) | 14.00 ± 0.00 (14) | 14 |
| | SC | 12.00 ± 0.00 (12) | 23.00 ± 0.00 (23) | 5.00 ± 0.00 (5) | 40.00 ± 0.00 (40) | 40 |
| | GES | 4.00 ± 0.00 (4) | 4.00 ± 0.00 (4) | 6.00 ± 0.00 (6) | 14.00 ± 0.00 (14) | 14 |
| | PC | 0.00 ± 0.00 (0) | 20.00 ± 0.00 (20) | 6.00 ± 0.00 (6) | 26.00 ± 0.00 (26) | 26 |
| | TPDA | 1.00 ± 0.00 (1) | 13.00 ± 0.00 (13) | 4.00 ± 0.00 (4) | 18.00 ± 0.00 (18) | 18 |
| | MMHC | 2.00 ± 0.00 (2) | 3.00 ± 0.00 (3) | 3.00 ± 0.00 (3) | 8.00 ± 0.00 (8) | 8 |
| Insurance-3000 | BFO-B | 3.10 ± 1.29 (1) | 0.25 ± 0.44 (0) | 3.80 ± 1.67 (1) | 7.15 ± 3.00 (2) | 9 |
| | ABC-B | 3.85 ± 1.18 (2) | 0.60 ± 0.50 (0) | 7.60 ± 3.51 (3) | 12.05 ± 4.81 (6) | 17 |
| | ACO-B | 2.95 ± 1.47 (1) | 0.45 ± 0.51 (0) | 5.75 ± 3.70 (1) | 9.15 ± 5.25 (2) | 14 |
| | GA-B | 3.80 ± 1.28 (2) | 0.85 ± 0.49 (0) | 3.65 ± 2.50 (1) | 8.30 ± 3.57 (4) | 10 |
| | HCST-B | 10.00 ± 0.11 (10) | 9.00 ± 0.00 (9) | 13 ± 0.00 (13) | 32.00 ± 0.00 (32) | 32 |
| | SC | 14.00 ± 0.00 (14) | 22.00 ± 0.00 (22) | 10.00 ± 0.00 (10) | 46.00 ± 0.00 (46) | 46 |
| | GES | 2.30 ± 0.00 (2) | 1.00 ± 0.00 (1) | 7.00 ± 0.00 (7) | 10.00 ± 0.00 (10) | 10 |
| | PC | 1.00 ± 0.00 (1) | 15.00 ± 0.00 (15) | 19.00 ± 0.00 (19) | 35.00 ± 0.00 (35) | 35 |
| | TPDA | 2.00 ± 0.00 (2) | 18.00 ± 0.00 (18) | 17.00 ± 0.00 (17) | 37.00 ± 0.00 (37) | 37 |
| | MMHC | 4.00 ± 0.00 (4) | 4.00 ± 0.00 (4) | 17.00 ± 0.00 (17) | 25.00 ± 0.00 (25) | 25 |

**Table 14.** (Continued)

| | | | | | |
|---|---|---|---|---|---|
| Child-2000 | BFO-B | 0.00 ± 0.00 (0) | 0.00 ± 0.00 (0) | 1.45 ± 1.07 (1) | 1.45 ± 1.07 (1) 1 |
| | ABC-B | 0.00 ± 0.00 (0) | 0.00 ± 0.00 (0) | 1.75 ± 1.30 (1) | 1.75 ± 1.30 (1) 1 |
| | ACO-B | 0.00 ± 0.00 (0) | 0.00 ± 0.00 (0) | 1.00 ± 0.00 (1) | 1.00 ± 0.00 (1) 1 |
| | GA-B | 0.00 ± 0.00 (0) | 0.00 ± 0.00 (0) | 1.00 ± 0.00 (1) | 1.00 ± 0.00 (1) 1 |
| | HCST-B | 0.00 ± 0.00 (0) | 0.00 ± 0.00 (0) | 3.00 ± 0.00 (3) | 3.00 ± 0.00 (3) 3 |
| | SC | 3.00 ± 0.00 (3) | 9.00 ± 0.00 (9) | 4.00 ± 0.00 (4) | 17.00 ± 0.00 (17) 17 |
| | GES | 0.00 ± 0.00 (0) | 5.00 ± 0.00 (5) | 7.00 ± 0.00 (7) | 12.00 ± 0.00 (12) 12 |
| | PC | 1.00 ± 0.00 (1) | 7.00 ± 0.00 (7) | 12.00 ± 0.00 (12) | 20.00 ± 0.00 (20) 20 |
| | TPDA | 1.00 ± 0.00 (1) | 6.00 ± 0.00 (6) | 14.00 ± 0.00 (14) | 21.00 ± 0.00 (21) 21 |
| | MMHC | 0.00 ± 0.00 (0) | 1.00 ± 0.00 (1) | 5.00 ± 0.00 (5) | 6.00 ± 0.00 (6) 6 |
| Asia-5000 | BFO-B | 0.00 ± 0.00 (0) | 0.00 ± 0.00 (0) | 1.00 ± 0.00 (1) | 1.00 ± 0.00 (1) 1 |
| | ABC-B | 0.00 ± 0.00 (0) | 0.00 ± 0.00 (0) | 1.00 ± 0.00 (1) | 1.00 ± 0.00 (1) 1 |
| | ACO-B | 0.00 ± 0.00 (0) | 0.00 ± 0.00 (0) | 1.00 ± 0.00 (1) | 1.00 ± 0.00 (1) 1 |
| | GA-B | 0.25 ± 0.55 (0) | 0.05 ± 0.22 (0) | 0.65 ± 1.03 (0) | 0.95 ± 1.64 (0) 1 |
| | HCST-B | 3.00 ± 0.00 (3) | 1.00 ± 0.00 (1) | 3.00 ± 0.00 (3) | 7.00 ± 0.00 (7) 7 |
| | SC | 0.00 ± 0.00 (0) | 1.00 ± 0.00 (1) | 1.00 ± 0.00 (1) | 2.00 ± 0.00 (2) 2 |
| | GES | 0.00 ± 0.00 (0) | 0.00 ± 0.00 (0) | 1.00 ± 0.00 (1) | 1.00 ± 0.00 (1) 1 |
| | PC | 0.00 ± 0.00 (0) | 6.00 ± 0.00 (6) | 1.00 ± 0.00 (1) | 7.00 ± 0.00 (7) 7 |
| | TPDA | 0.00 ± 0.00 (0) | 4.00 ± 0.00 (4) | 1.00 ± 0.00 (1) | 5.00 ± 0.00 (5) 5 |
| | MMHC | 0.00 ± 0.00 (0) | 0.00 ± 0.00 (0) | 2.00 ± 0.00 (2) | 2.00 ± 0.00 (2) 2 |

**Table 15.** Structural differences among 10 algorithms on Studfarm-10000, Boerlage92-10000, Brain-10000, and Hepar II-50000.

| Datasets | Algorithms | AAD | ADD | ARD | ASD | SLQ |
|---|---|---|---|---|---|---|
| Studfarm-10000 | BFO-B | 0.00 ± 0.00 (0) | 0.00 ± 0.00 (0) | 0.00 ± 0.00 (0) | 0.00 ± 0.00 (0) | 0 |
| | ABC-B | 0.00 ± 0.00 (0) | 0.00 ± 0.00 (0) | 0.00 ± 0.00 (0) | 0.00 ± 0.00 (0) | 0 |
| | ACO-B | 0.00 ± 0.00 (0) | 0.00 ± 0.00 (0) | 0.00 ± 0.00 (0) | 0.00 ± 0.00 (0) | 0 |
| | GA-B | 0.80 ± 0.41 (0) | 0.00 ± 0.00 (0) | 0.95 ± 0.60 (0) | 1.75 ± 0.97 (0) | 2 |
| | HCST-B | 6.00 ± 0.00 (6) | 0.00 ± 0.00 (0) | 10.00 ± 0.00 (10) | 16.00 ± 0.00 (16) | 16 |
| | SC | 5.00 ± 0.00 (5) | 4.00 ± 0.00 (4) | 3.00 ± 0.00 (3) | 12.00 ± 0.00 (12) | 12 |
| | GES | 7.00 ± 0.00 (7) | 2.00 ± 0.00 (2) | 2.00 ± 0.00 (2) | 11.00 ± 0.00 (11) | 11 |
| | PC | 0.00 ± 0.00 (0) | 6.00 ± 0.00 (6) | 2.00 ± 0.00 (2) | 8.00 ± 0.00 (8) | 8 |
| | TPDA | 0.00 ± 0.00 (0) | 3.00 ± 0.00 (3) | 1.00 ± 0.00 (1) | 4.00 ± 0.00 (4) | 4 |
| | MMHC | 1.00 ± 0.00 (1) | 0.00 ± 0.00 (0) | 0.00 ± 0.00 (0) | 1.00 ± 0.00 (1) | 1 |
| Boerlage92-10000 | BFO-B | 0.45 ± 0.51 (0) | 6.90 ± 0.64 (6) | 5.30 ± 0.47 (5) | 12.65 ± 1.31 (11) | 14 |
| | ABC-B | 0.95 ± 0.22 (0) | 7.30 ± 0.47 (7) | 5.65 ± 0.49 (5) | 13.90 ± 0.45 (12) | 14 |
| | ACO-B | 1.10 ± 0.45 (0) | 7.00 ± 0.00 (7) | 7.30 ± 3.33 (5) | 15.40 ± 3.73 (12) | 14 |
| | GA-B | 1.30 ± 0.64 (1) | 7.20 ± 0.41 (7) | 9.55 ± 1.79 (6) | 18.65 ± 2.30 (14) | 19 |
| | HCST-B | 4.00 ± 0.00 (4) | 6.00 ± 0.00 (6) | 17.00 ± 0.00 (17) | 27.00 ± 0.00 (27) | 27 |
| | SC | 13.00 ± 0.00 (13) | 26.00 ± 0.00 (26) | 4.00 ± 0.00 (4) | 43.00 ± 0.00 (43) | 42 |
| | GES | 3.00 ± 0.00 (3) | 11.00 ± 0.00 (11) | 8.00 ± 0.00 (8) | 22.00 ± 0.00 (22) | 22 |
| | PC | 0.00 ± 0.00 (0) | 27.00 ± 0.00 (27) | 6.00 ± 0.00 (6) | 33.00 ± 0.00 (33) | 33 |
| | TPDA | 0.00 ± 0.00 (0) | 24.00 ± 0.00 (24) | 7.00 ± 0.00 (7) | 31.00 ± 0.00 (31) | 31 |
| | MMHC | 0.00 ± 0.00 (0) | 8.00 ± 0.00 (8) | 6.00 ± 0.00 (6) | 14.00 ± 0.00 (14) | 14 |

## Table 15. (Continued)

| | | | | | |
|---|---|---|---|---|---|
| Brain-10000 | BFO-B | **0.00 ± 0.00 (0)** | **0.00 ± 0.00 (0)** | 17.05 ± 2.96 (14) | **17.05 ± 2.96 (14)** | 19 |
| | ABC-B | **0.00 ± 0.00 (0)** | **0.00 ± 0.00 (0)** | 19.40 ± 2.96 (15) | 19.40 ± 2.96 (15) | 21 |
| | ACO-B | **0.00 ± 0.00 (0)** | **0.00 ± 0.00 (0)** | 17.90 ± 2.88 (**13**) | 17.90 ± 2.88 (**13**) | 20 |
| | GA-B | 2.50 ± 2.24 (0) | **0.00 ± 0.00 (0)** | **15.75 ± 3.19 (13)** | 18.25 ± 5.30 (14) | **15** |
| | HCST-B | **0.00 ± 0.00 (0)** | **0.00 ± 0.00 (0)** | 32.00 ± 0.00 (32) | 32.00 ± 0.00 (32) | 32 |
| | SC | 39.00 ± 0.00 (39) | 4.00 ± 0.00 (4) | 29.00 ± 0.00 (29) | 72.00 ± 0.00 (72) | 72 |
| | GES | **0.00 ± 0.00 (0)** | 10.00 ± 0.00 (10) | 28.00 ± 0.00 (28) | 38.00 ± 0.00 (38) | 38 |
| | PC | 732.00 ± 0.00 (732) | **0.00 ± 0.00 (0)** | 58.00 ± 0.00 (58) | 790.00 ± 0.00 (790) | 790 |
| | TPDA | 29.00 ± 0.00 (29) | 60.00 ± 0.00 (60) | 1.00 ± 0.00 (1) | 90.00 ± 0.00 (90) | 90 |
| | MMHC | 90.00 ± 0.00 (90) | 7.00 ± 0.00 (7) | 43.00 ± 0.00 (43) | 140.00 ± 0.00 (140) | 140 |
| Hepar II-50000 | BFO-B | 6.20 ± 1.44 (5) | 11.70 ± 0.57 (10) | 10.75 ± 2.15 (9) | **28.65 ± 3.34 (26)** | **31** |
| | ABC-B | 8.70 ± 1.56 (3) | **10.15 ± 0.88 (9)** | 20.10 ± 4.84 (8) | 38.95 ± 5.46 (23) | 41 |
| | ACO-B | 5.85 ± 1.93 (3) | 10.95 ± 0.83 (9) | 13.60 ± 3.65 (9) | 30.40 ± 4.31 (26) | **31** |
| | GA-B | 8.25 ± 3.90 (3) | 10.75 ± 1.30 (10) | 15.75 ± 5.36 (9) | 34.75 ± 8.84 (**22**) | 43 |
| | HCST-B | 2.00 ± 0.00 (2) | 13.00 ± 0.00 (13) | 18.00 ± 0.00 (18) | 33.00 ± 0.00 (33) | 33 |
| | SC | 36.00 ± 0.00 (36) | 116.00 ± 0.00 (116) | **3.00 ± 0.00 (3)** | 155.00 ± 0.00 (155) | 155 |
| | GES | 1.00 ± 0.00 (1) | 31.00 ± 0.00 (31) | 9.00 ± 0.00 (9) | 41.00 ± 0.00 (41) | 41 |
| | PC | **0.00 ± 0.00 (0)** | 110.00 ± 0.00 (110) | 9.00 ± 0.00 (9) | 119.00 ± 0.00 (119) | 119 |
| | TPDA | **0.00 ± 0.00 (0)** | 103.00 ± 0.00 (103) | 14.00 ± 0.00 (14) | 117.00 ± 0.00 (117) | 117 |
| | MMHC | 3.00 ± 0.00 (3) | 28.00 ± 0.00 (28) | 7.00 ± 0.00 (7) | 38.00 ± 0.00 (38) | 38 |

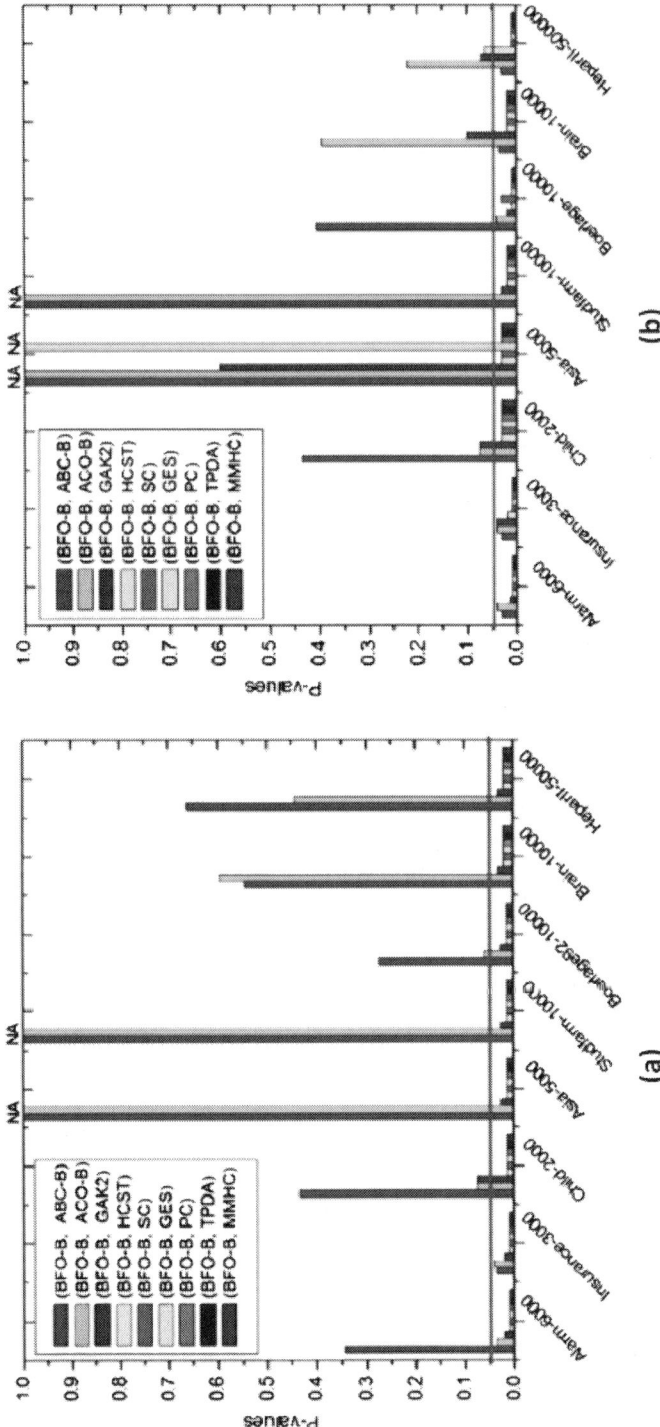

**Figure 4.** Pair-wise Kruskal–Wallis tests: (a) K2 scores, (b) structure.

sults of two algorithms are exactly the same over all runs, $p$ is infinite, and we denote this case with "NA" in Fig. 4.

**(A) K2 scores** The three swarm intelligence based algorithms, BFO-B, ABC-B, and ACO-B, obtain the highest scores (in bold) with respect to the four metrics (HKS, LKS, KLQ, and AKS) on all eight datasets (Table 12 and Table 13), which illustrates the superiority of swarm intelligence based approaches over score and search approaches used by the other algorithms. A scoring metric was used to measure the degree of matching between the learned network and the dataset. Thus, in spite of not using the scoring mechanism for the PC and TPDA algorithms, we are still able to measure their outcomes as inferior to the three swarm intelligence algorithms. The reasons for the inferior obtained by the GA-B, HCST-B, SC, GES, MMHC, PC, and TPDA algorithms are

- Although GA-B is a global optimization method, it searches the node ordering using a genetic algorithm, which is easily trapped into local optima, especially when the node number is large.
- Although the HCST-B, SC, GES and MMHC algorithms have different concrete implementations, they all look for better networks using a greedy hill climbing search, which is a local optimization technology, and can only ensure these algorithms find local optimal solutions.
- PC and TPDA make use of statistical or information theory measures to judge whether certain arcs between variables exist. This type of method is unreliable for finding better networks.

For the three swarm intelligence algorithms (BFO-B, ABC-B, and ACO-B), the proposed BFO-B algorithm slightly outperforms ABC-B and ACO-B, and obtains the highest AKS on five of the eight datasets (Alarm-6000, Insurance-3000, Asia-5000, Studfarm-10000, and Boerlage92-10000).

Fig. 4(a) shows the BFO-B algorithm is significantly superior to the seven non-swarm intelligence algorithms (GA-B, HCST-B, SC, GES, PC, TPDA, and MMHC) on all eight datasets (except GA-B on Child-2000), with higher K2 scores and $p<0.05$ (except for BFO-B and GA-B on Child-2000). BFO-B performs significantly better than ABC-B on Insurance-3000 and ACO-B on Alarm-6000 and Insurance-3000 ($p<0.05$).

However, there are no significant differences between BFO-B and either of ABC-B and ACO-B for the other cases.

Hence, the proposed BFO-B algorithm can guarantee the discovery of better quality networks, with higher scores than the seven non-swarm intelligence algorithms (GA-B, HCST-B, SC, GES, PC, TPDA, and MMHC), and are not lower than the two swarm intelligence algorithms, ABC-B and ACO-B.

**(B) Structure differences** The results in terms of structure difference are similar to those in terms of scores (Table 14 and Table 15). The swarm intelligence algorithms (BFO-B, ABC-B, and ACO-B) have smaller structural differences than the other seven algorithms, GA-B, HCST-B, SC, GES, PC, TPDA, and MMHC, according to the structural difference metrics (AAD, ADD, ARD, ASD, and SLQ) in the majority of situations. The proposed BFO-B algorithm obtains the smallest ASD on six of the eight datasets (Alarm-6000, Insurance-3000, Studfarm-10000, Boerlage92-10000, Brain-10000, and Hepar II-50000).

There are several situations where the structural differences are too large. For example, the MMHC and PC algorithms are respectively 140 and 790 for the Brain-10000 data, and the SC, PC and TPDA algorithms are respectively 155, 119, and 117 for the Hepar II-50000 data. These results indicate that the corresponding algorithms completely failed to learn the network structures from the datasets. Two factors may be the cause of such poor outcomes: the default parameters are potentially unsuitable for the corresponding algorithms on these datasets (Brain-10000 and Hepar II-50000); and/or the networks (Brain and Hepar II) may be more complex, and the number of cases (10000 and 50000) are not large enough for the corresponding algorithms to correctly learn better network structures.

Fig. 4(b) shows that the BFO-B algorithm is significantly superior to HCST, SC, GES, PC, TPDA, and MMHC, on all eight datasets (except GES on Asia-5000 and HCST-B on Hepar II-50000), achieving the smallest ASD, and the corresponding $p$-values are all less than 0.05. BFO-B also performs significantly better than GA-B on half of the datasets (Alarm-6000, Insurance-3000, Studfarm-10000, and Boerlage92-10000), ABC-B also on (a different) half of the datasets (Alarm-6000, Insurance-3000, Brain-10000, and Hepar II-50000), and ACO-B on three datasets (Alarm-6000, Insurance-3000, and Boerlage92-10000). For the other cases, there are no significant differences between BFO-B and GA-B, ABO-B, and ACO-B. Hence,

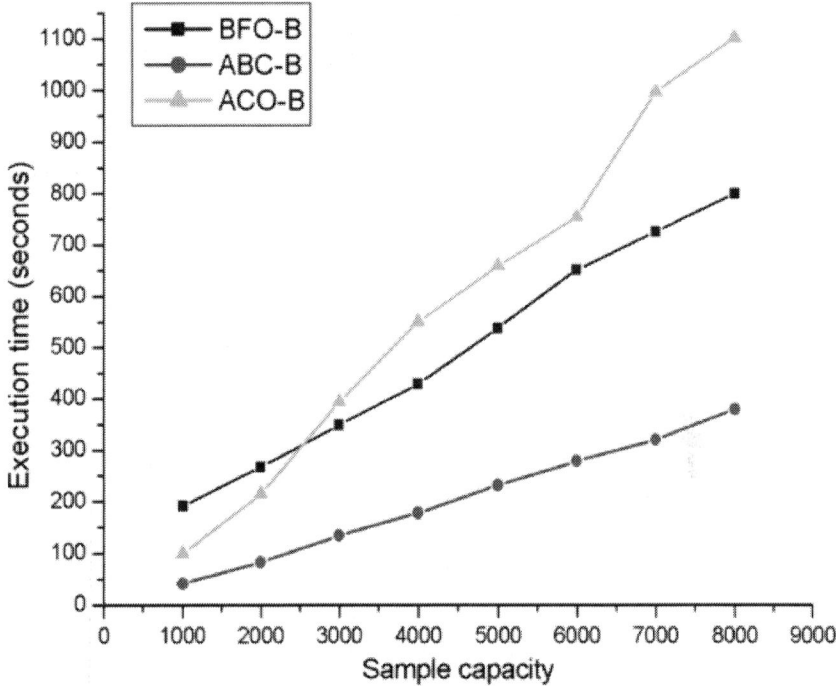

**Figure 5.** Time performance for three algorithms on the Alarm network.

the proposed BFO-B algorithm can obtain networks with smaller structural differences compared with the other algorithms.

**(C) Time performance** Table 12 and Table 13 show that the four population based algorithms BFO-B, ABC-B, ACO-B, and GA-B generally take more time than the non-population based algorithms (HCST-B, SC, GES, PC, TPDA, and MMHC), because these four algorithms have many candidate solutions to be optimized. The GA-B algorithm also needs more time than the three swarm intelligence algorithms BFO-B, ABC-B, and ACO-B, because it performs both genetic and K2 algorithms each iteration.

To further study the time performance of BFO-B among swarm intelligence based algorithms, we compared the three swarm intelligence algorithms on eight datasets generated from the Alarm network, as shown in Fig. 5, where the execution time is an average of 20 independent runs. BFO-B performs the worst among the three algorithms on Alarm-2000 and Alarm-3000. This is because the solution initialization and optimization phases are separated, in contrast to ABC-B and ACO-B, which merge optimization processes into their solution construction phases. Thus, when the sample capacity is small, BFO-B spends a relatively high proportion of its time on initialization, which leads to an overall longer running time. However, as the sample capacity increases, BFO-B spends most of its time on optimization, and initialization time becomes trivial, as shown in Fig. 5. BFO-B performance improves, and time performance stays between that of ABC-B and ACO-B when the sample capacity equals or exceeds 3000. However, because ABC-B has two information exchange mechanisms among artificial bees, it can quickly perform solution optimization and obtains the best time performance.

Summarizing, we conclude that the proposed BFO-B algorithm is capable of finding networks that is superior to those obtained by the non-swarm intelligence algorithms (GA-B, HCST-B, SC, GES, PC, TPDA, and MMHC) and is highly competitive compared with networks obtained by the two state of the art swarm intelligence based algorithms (ABC-B and ACO-B). Therefore, the proposed BFO-B algorithm appears to be a promising method for learning BN structures from data.

## 5. CONCLUSIONS

We proposed a new swarm intelligence algorithm, BFO-B, to learn the structures of BNs. The proposed algorithm uses BFO to search the optimal network structure in the network space and the K2 metric to guide the search process. The novelty of this method is its application of basic BFO to the structural learning of BNs, which not only enriches the application of the BFO B algorithm but also provides a new way to learn BN structure in uncertain artificial intelligence. We performed a series of experiments to demonstrate the performance of the BFO-B algorithm in many domains and compared the algorithm with nine representative methods: ACO-B, ABC-B, GA-B, HCST-B, SC, GES, PC, TPDA, and MMHC. The proposed algorithm BFO-B algorithm is an effective algorithm to accurately learn a BN structure from the data.

In future work, we will continue to study the optimization mechanisms of BFO to improve BFO-B time performance, and extend our study to more complex optimization problems, such as dynamic BN learning problems and BN learning problems with incomplete data or large scale nodes.

## ACKNOWLEDGEMENTS

This work was partly supported by the NSFC Research Program (61375059, 61332016), the National "973" Key Basic Research Program of China (2014CB744601), the Specialized Research Fund for the Doctoral Program of Higher Education (20121103110031), and the Beijing Municipal Education Research Plan key project (Beijing Municipal Fund Class B) (KZ201410005004).

## REFERENCES

1. J. Pearl, Reasoning in Intelligent Systems: Networks of Plausible Inference, Morgan Kaufmann, San Mateo, CA, 1988.
2. R. Daly, Q. Shen, S. Aitken, Learning Bayesian networks: approaches and issues, Knowl. Eng. Rev. 26 (2) (2011) 99–157.
3. L.M. de Campos, J.M. Fernández-Luna, J.M. Puerta, Iterated local search algorithm for learning Bayesian networks with restarts on conditional indepen- dence tests, Int. J. Intell. Syst. 18 (2) (2003) 221–235.
4. M.L. Wong, K.S. Leung, Using evolutionary programming and minimum description length principle for data mining of Bayesian networks, IEEE Trans. Pattern Anal. Mach. Intell. 21 (2) (1999) 174–178.
5. M.L. Wong, K.S. Leung, An e□cient data mining method for learning Bayesian network using an evolutionary algorithm-based hybrid approach, IEEE Trans. Evol. Comput. 8 (4) (2004) 378–404.
6. L.M. de Campos, J.M. Fernández-Luna, J.A. Gámez, J.M. Puerta, Ant colony optimization for learning Bayesian networks, Int. J. Approx. Reason. 31 (3) (2002) 291–311.
7. L.M. de Campos, J.F. Hutte, A new approach for learning belief networks using independence criteria, Int. J. Approx. Reason. 24 (1) (2000) 11–37.
8. P. Larrañaga, M. Poza, Y. Yurramendi, R.H. Murga, C.M.H. Kuijpers, Structure learning of Bayesian networks by genetic algorithms: a performance analysis of control parameters, IEEE Trans. Pattern Anal. Mach. Intell. 18 (9) (1996) 912–925.
9. P. Larrañaga, C.M.H. Kuijpers, R.H. Murga, Y. Yurramendi, Learning Bayesian network structures by searching for the best ordering with

genetic algo- rithms, IEEE Trans. Syst. Man Cybern., Part A, Syst. Hum. 26 (4) (1996) 487–493.
10. J.Z. Ji, H.K. Wei, C.N. Liu, An artificial bee colony algorithm for learning Bayesian networks, Soft Comput. 17 (6) (2013) 983–984.
11. J.Z. Ji, R.B. Hu, H.X. Zhang, C.N. Liu, A hybrid method for learning Bayesian networks based on ant colony optimization, Appl. Soft Comput. 11 (4) (2011) 3373–3384.
12. J.Z. Ji, H.X. Zhang, R.B. Hu, C.N. Liu, A Bayesian network learning algorithm based on independence test and ant colony optimization, Acta Autom. Sin.35 (3) (2009) 281–288.
13. J.Z. Ji, H.X. Zhang, R.B. Hu, C.N. Liu, A Tabu-search based Bayesian network structural learning algorithm, J. Beijing Univ. Technol. 37 (8) (2011) 1274–1280.
14. J.A. Gámez, J.M. Puerta, Searching for the best elimination sequence in Bayesian networks by using ant colony optimization, Pattern Recognit. Lett. 23 (1–3) (2002) 261–277.
15. I. Tsamardinos, L.E. Brown, C.F. Aliferis, The Max–Min Hill-Climbing Bayesian network structural learning algorithm, Mach. Learn. 65 (1) (2006) 31–78.
16. J. Suzuki, Learning Bayesian belief networks based on the minimum description length principle: basic properties, IEICE Trans. Fundam. Electron.Commun. Comput. Sci. 82 (10) (1999) 2237–2245.
17. D.Y. Liu, F. Wang, Y.N. Lu, W.X. Xue, S.X. Wang, Research on learning Bayesian network structure based on genetic algorithms, Comput. Sci. Res. Dev.38 (8) (2001) 916–922.
18. P.C. Pinto, A. Nägele, M. Dejori, T.A. Runkler, J.M.C. Sousa, Using a local discovery ant Algorithm for Bayesian network structural learning, IEEE Trans.Evol. Comput. 13 (4) (2009) 767–778.
19. J. Cheng, D.A. Bell, W. Liu, Learning belief networks from data: an information theory based approach, in: Proceedings of the 6th International Confer- ence on Information and Knowledge Management, CIKM'97, 1997, pp. 325–331.
20. D. Heckerman, A tutorial on learning with Bayesian networks, in: Innovations in Bayesian Networks – Studies in Computational Intelligence, Springer, Berlin, Heidelberg, 2008, pp. 33–82.
21. N. Friedman, I. Nachman, D. Peér, Learning Bayesian network structure from massive datasets: the "sparse candidata" algorithm, in: Pro-

ceedings of the 15th Conference on Uncertainty in Artificial Intelligence, UAI'99, 1999, pp. 206–215.
22. H.X. Chen, Q. Zheng, T. Lei, S.L. Ping, Research on learning Bayesian networks by particle swarm optimization, Inf. Technol. J. 5 (3) (2006) 540–545.
23. H.X. Chen, Q. Zheng, T. Lei, S.L. Ping, Learning Bayesian network structures with discrete particle swarm optimization algorithm, in: Proceedings of the 2007 IEEE Symposium on Foundations of Computational Intelligence, FOCI2007, 2007, pp. 47–52.
24. T. Wang, J. Yang, A heuristic method for learning Bayesian networks using discrete particle swarm optimization, Knowl. Inf. Syst. 24 (2) (2010) 269–281.
25. G.F. Cooper, E. Herskovits, A Bayesian method for the induction of probabilistic networks from data, Mach. Learn. 9 (4) (1992) 309–347.
26. D. Heckerman, D. Geiger, D.M. Chickering, Learning Bayesian networks: the combination of knowledge and statistical data, Mach. Learn. 20 (3) (1995) 197–243.
27. J.R. Alcobé, Incremental hill-climbing search applied to Bayesian network structural learning, in: Proceedings of the 15th European Conference on Machine Learning, Pisa, Italy, 2004.
28. D.M. Chickering, Optimal structure identification with greedy search, J. Mach. Learn. Res. 3 (2002) 507–554.
29. P. Spirtes, C. Glymour, R. Scheines, Causation, Prediction, and Search, second edition, MIT Press, 2002.
30. D.M. Chickering, D. Geiger, D. Heckerman, Learning Bayesian networks is NP-Hard, Technical Report MSR-TR-94-17, Microsoft Research, 1994.
31. Y.H. Shi, An optimization algorithm based on brainstorming process, Int. J. Swarm Intel. Res. 2 (4) (2011) 35–62.
32. J. Kennedy, R. Eberhart, Particle swarm optimization, in: Proceedings of the 1995 IEEE International Conference on Neural Networks, 1995, pp. 1942–1948.
33. Y.H. Shi, R. Eberhart, A modified particle swarm optimizer, in: Proceedings of the 1998 IEEE World Congress on Computational Intelligence, 1998, pp. 69–73.
34. M. Dorigo, V.A. Maniezzo, A. Colorni, The ant system: optimization by a colony of cooperating agents, IEEE Trans. Syst. Man Cybern., Part B, Cybern. 26 (1) (1996) 29–41.

35. M. Dorigo, L.M. Gambardella, Ant colony system: a cooperative learning approach to the traveling salesman problem, IEEE Trans. Evol. Comput. 1 (1) (1997) 53–66.
36. D. Karaboga, An idea based on honey bee swarm for numerical optimization, Technical Report TR06, Computer Engineering Department, Erciyes University, Turkey, 2005.
37. K.M. Passino, Biomimicry of bacterial foraging for distributed optimization and control, IEEE Control Syst. Mag. 22 (3) (2002) 52–67.
38. Y. Liu, K.M. Passino, Biomimicry of social foraging bacteria for distributed optimization: models, principles, and emergent behaviors, J. Optim. Theory Appl. 115 (3) (2002) 603–628.
39. M. Wan, L.X. Li, J.H. Xiao, C. Wang, Y.X. Yang, Data clustering using bacterial foraging optimization, J. Intell. Inf. Syst. 38 (2) (2012) 321–341.
40. V. Agrawal, H. Sharma, J.C. Bansal, Bacterial foraging optimization: a survey, Adv. Intell. Soft Comput. 130 (2012) 227–242.
41. S. Mishra, A hybrid least square-fuzzy bacterial foraging strategy for harmonic estimation, IEEE Trans. Evol. Comput. 9 (1) (2005) 61–73.
42. S. Das, A. Biswas, S. Dasgupta, A. Abraham, Bacterial foraging optimization algorithm: theoretical foundations, analysis, and applications, Found. Com- put. Intel. 3 (2009) 23–55.
43. O.P. Verma, M. Hanmandlu, P. Kumar, S. Chhabra, A. Jindal, A novel bacterial foraging technique for edge detection, Pattern Recognit. Lett. 32 (2011) 1187–1196.
44. H.N. Chen, Y.L. Zhu, K.Y. Hu, Multi-colony bacteria foraging optimization with cell-to-cell communication for RFID network planning, Appl. Soft Comput. 10 (2010) 539–547.
45. A. Abraham, A. Biswas, S. Dasgupta, S. Das, Analysis of reproduction operator in bacterial foraging optimization algorithm, IEEE Trans. Evol. Comput. (2008) 1476–1483.
46. Á. Rubio-Largo, M.A. Vega-Rodríguez, J.A. Gómez-Pulido, J.M. Sánchez-Pérez, A comparative study on multiobjective swarm intelligence for the routing and wavelength assignment problem, IEEE Trans. Syst. Man Cybern., Part C, Appl. Rev. 42 (6) (2012) 1644–1655.

# INDEX

3rd Generation Partnership Project  166

## A

aluminum alloy  141
ASTF  29

## B

Bayesian Belief Network  40
Bayesian Markov switching method  63
Bayes' theorem  17
beta-binomial model  196
BFO  248
binomial distribution  2

## C

Cellular LTE downlink  187
CGMM  110
Chemotaxis  249, 253
Cointegrated Price Series Analysis  72
Cointegration  64
Condition Diagnosis by NN  56
covariance matrix  16

## D

Decision on Belief  2
Delaunay triangulation  102
Dynamic Programming  177

## E

Engle-Granger method  73
evaluation factor  30

## F

fault detection and diagnosis  16
Feature Extraction  32
Fourier analysis  30

## G

Gaussians  83
geospatial approach  94
Grassmann manifold  132

## H

heuristic threshold policy  16
High Sensitivity Symptom Parameters  40
Hilbert-Huang transform  31
HKS  263
homoscedasticity  196, 200

## I

ICP signal  143
Impact-Echo Signals  129

## K

KL-divergence  203

## L

lane-level information 94
Laplace approximation 207
Long-Term Evolution 165

## M

Markov Blanket 85
Markov chain Monte Carlo 196
MCMC sampling 213
Meta-heuristic 87
MLIT 93, 123
Model Assessment 68

## N

Naïve Based Bayesian 77
Naïve Bayesian Classifier 118
NDT: sensors 130
neuroimaging data 192
nonstationary time series 64
normal-binomial model 196

## O

OFDM 168
OOCOOC 16
OpenStreetMap 97

## P

pathophysiological states 191
Posterior Decision Making 69
posteriori 130, 156
P-wave 131

## Q

Qos Attributes 80
QPSK modulation scheme 177
QWS dataset 79

## R

Resource Block 165
Road Plane Extraction 107
Roc Curves 156

## S

semi-Markov models 74
semi-supervised scenario 133
Simulated Time Series Analysis 68
subject-specific accuracies 219
S-wave 131

## T

Tabu search 77
Tdp Technique 169
trajectory segment section 95
TSW 114

## V

Variational inference 202

## W

Web Services 77